The Unknown World of the Mobile Home

Creating the North American Landscape

Gregory Conniff
Edward K. Muller
David Schuyler
Consulting Editors

George F. Thompson
Series Founder and Director

Published in cooperation with
the Center for American Places,
Santa Fe, New Mexico, and
Harrisonburg, Virginia

THE UNKNOWN WORLD OF THE MOBILE HOME

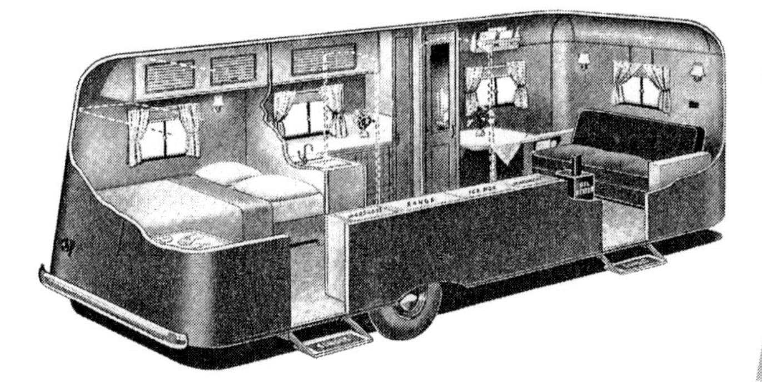

John Fraser Hart,

Michelle J. Rhodes,

and John T. Morgan,

with the cartographic

collaboration of

Mark B. Lindberg

The Johns Hopkins

University Press

Baltimore and London

Printed in the United States of America on acid-free paper

9 8 7 6 5 4 3 2 1

The Johns Hopkins University Press

2715 North Charles Street
Baltimore, Maryland 21218-4363
www.press.jhu.edu

Library of Congress Cataloging-in-Publication Data
Hart, John Fraser.
The unknown world of the mobile home / John Fraser Hart, Michelle J. Rhodes and John T. Morgan.
p. cm. — (Creating the North American landscape)
Includes bibliographical references and index.
ISBN 0-8018-6899-8 (alk. paper)
1. Mobile homes—United States. 2. Mobile home living—United States. 3. Mobile home industry—United States. 4. Mobile home parks—United States. I. Rhodes, Michelle J., 1972– II. Morgan, John (John T.) III. Title. IV. Series.
HD7289.62.U6 H37 2002
307.3′36—dc21
2001004067

A catalog record for this book is available from the British Library.

To the memory of John Brinckerhoff Jackson

CONTENTS

ACKNOWLEDGMENTS

All of the photographs in this book, unless otherwise stipulated, are by one of the authors. Mark Lindberg drew all the maps.

We are delighted to express our appreciation for the help we have received from John Adams, Erin Arnold, Therese Bissell, Al Carlson, Len Evenden, Jon Everitt, Tom Field, Dick Foster, Helen Hallis, Bob Horsfall, Brinck Jackson, Kathy Johnson, Linda and Steve Johnson, Sid Jumper, Kathy Klink, Joe Krasne, Jodi Larson, Heather Lawson, Rob Martin, Chris Mayda, Bob Mings, Ruth Nelson, Richard Pillsbury, Karl Raitz, Margaret Rasmussen, Bard Rupp, Pat Smith, Rob Tunnell, John Upchurch, Connie Weil, Jeff Wick, and all of those we have quoted and cited in the text.

We especially appreciate the support and encouragement of George F. Thompson, president of the Center for American Places, who has done more than any other person to encourage geographers to write books.

We are grateful to Mary V. Yates for her splendid job of copyediting and to Carol Zimmerman, who has shepherded our manuscript through production.

The Unknown World of the Mobile Home

PART ONE

BACKGROUND

Prologue

Like the old gray mare, mobile homes and trailers ain't what they used to be, or what many people still think they are. They are wider, longer, and vastly better. The 8-by-40-foot trailer of the 1950s evolved into the 14-by-70-foot mobile home of the 1990s. Some are still distinctive and easily identifiable rectangular boxes, with flat roofs and gleaming metal siding, but many have pitched roofs and vinyl siding, and the new multisectional double-wides, which consist of two halves that are transported separately, joined at the site, and never moved thereafter, are virtually indistinguishable from conventional site-built houses.

The demand for mobile homes seems sure to increase, because the average American family can no longer afford the average price of a conventional site-built house. In the year 2000, mobile homes accounted for about 20 percent of all new single-family housing starts and about 30 percent of all new single-family homes sold, and enthusiasts predicted that mobile homes would soon comprise more than half of all new homes.

In 1993 Jeff Wick, president of Wick Building Systems, a leading manufacturer of mobile homes, selected a random sample of 50 people who had bought homes made by his company in the preceding year, and interviewed each one by telephone. Eighty-seven percent had lived in another mobile home previously, which he took as a sign that the industry was not doing a good job of getting the word out to potential new customers. Every person with whom he talked told him that first-time visitors to their new homes expressed surprise at how nice they were.

The 1990 Census of Housing reported that 7 of every 100 Americans lived in a mobile home. Most of the other 93 know precious little about this distinctive type of housing, and much of what they think they know is woefully wrong. Ignorance begets prejudice. Many Americans simply pretend that mobile homes do not exist, and if they think about them at all, which is not very often, they perceive mobile homes as cheap, flimsy, and undesirable housing for unattractive people. They assume that the residents of mobile homes are seriously deficient: deficient in income, deficient in education, deficient in intelligence, and deficient in moral fiber.

Many people think that mobile home parks depress the value of adjacent properties and increase traffic and crime. They are widely perceived as hotbeds of sex and violence, and the media are all too happy to pander to this perception. On the scale of general social acceptability, mobile home parks rank somewhere in the neighborhood of junkyards, but junkyards for people rather than for automobiles. They are segregated to remote and unattractive places in the less desirable parts of the outer urban fringes, discreetly distanced from other kinds of residential areas, and carefully camouflaged by plantings, fences, or high earthen balks to keep them from offending the sensitivities of better folk who might happen to wander past.

Early trailers earned their unsavory reputation for shoddy construction, inferior materials, and careless workmanship. They were susceptible to damage by high winds and might be twisted or blown over if not securely anchored. Their long narrow shape acted as a flue that channeled fires, and their layout could trap occupants in their bedrooms. They lacked adequate storage space and privacy, and suffered wide internal temperature gradients from floor to ceiling and from walls to room center. They depreciated more rapidly than site-built houses.

Congress addressed these problems with the Manufactured Home Construction and Safety Standards Act of 1974, which took effect in 1976. This act directed the Department of Housing and Urban Development (HUD) to develop a set of minimal quality and safety standards known as the HUD Code. The federal standards facilitate the marketing of standard models nationwide to buyers who can be reasonably sure of what they are getting, and they preempt a bewildering variety of local building codes that were designed to exclude mobile homes, but polities can still exclude mobile homes with regulations controlling such things as their design, roof pitch, siding, footage, and spacing.

Fig. 1. Mobile home park near the Orlando, Florida, airport

Improvements spawned name inflation. The first "trailers" were little more than wooden tents on wheels that were hauled behind the family car. They got a bad name when people started using them as semipermanent residences. In the 1950s manufacturers began to assert that their new, improved models were really "mobile homes." Few mobile homes are truly mobile, however; some people claim that 95 percent are never moved after they have been placed on their first site. The industry lobbied to change the name to "manufactured housing," and the 1980 Housing Act stipulated that "the term mobile home be changed to manufactured housing in all federal law and literature."

This stipulation has been generally ignored, and probably wisely, because the term "manufactured housing" is unfamiliar and confusing to people outside the industry. It includes precut or shell houses, panelized houses, modular or sectional houses, log houses, and geodesic dome houses as well as mobile homes, and thus we have elected to stick to the less confusing "mobile home." In vernacular usage it appears that "trailers" are single-wides that antedate the 1976 HUD Code, single-wides and double-wides are called "mobile homes," and "manufactured housing" is generally associated with houselike triples and quads. There is no clear and consistent nomenclature, and we have used the names that people used when we talked to them.

Modular homes are assembled on site by piecing together major components shipped from factories. Conventional homes once were built on site

Fig. 2. A brick foundation and brick pillars supporting the carport suggest that the owner of this double-wide "mobile" home does not plan to move it any time soon

Fig. 3. Mobile home in transit to Evergreen, Montana

from scratch, but nowadays their doors, windows, cabinets, and other minor components may also be assembled in factories and shipped prebuilt. Conventional homes are built and finished by skilled artisans, such as carpenters, brick masons, plumbers, painters, and electricians, all of whom receive union wages. Their work is subject to the vicissitudes of wind and weather, and the construction of conventional homes can be hampered by rain, extremes of temperature, wind, and blown dust.

Mobile homes are built completely under cover, except for their foundations, and they arrive on site ready for immediate occupancy. They are built on assembly lines in factories by semiskilled workers. Their manufacturers are able to achieve economies of scale by mass production, and they pay lower wages. In 1997 the Manufactured Housing Institute estimated that the average cost of manufactured homes was $25.78 per square foot for single-wides and $30.65 per square foot for double-wides, while the average cost of conventional site-built homes was $61.47 per square foot.

Mobile homes must be transported from the factory to their site, and there is risk of damage in transit and siting. A unit that is not properly supported and leveled on site can be seriously damaged, leaving it with ill-fitting doors and windows, cracked walls or ceilings, and buckled floors. It can flex or twist, causing poor window and door alignment and structural damage. Siting them has become more expensive and more complicated as they have become larger, heavier, and more like conventional site-built houses.

Even skeptical observers must admit that contemporary mobile homes are a highly acceptable and inexpensive alternative to conventional site-built houses. They are inexpensive starter homes for young couples and easily maintained retirement homes for elderly people. They cost less than half as much as site-built houses of comparable size. They have an average life expectancy of 20 to 35 years, as compared with 100 years for site-built houses, but many older mobiles are still in use. The normal mortgage on a mobile home is for only 7 to 12 years, in contrast to 20 to 25 years for a site-built house, and the interest rate is appreciably higher.

A crazy quilt of local regulations and zoning ordinances can hinder the siting of mobile homes in places where they would be welcome additions to the stock of affordable housing, and their ambivalent legal status complicates their regulation and taxation. Are they vehicles, or are they houses? Are they personal property, or are they real property? They are built, sold, and financed like automobiles, and traditionally they have been regulated and taxed as vehicles, because they have been presumed to be mobile, but today the vast majority are permanent dwelling units.

A mobile home parked on a lot owned by someone else can be considered movable personal property, but placing one on a permanent foundation on a site owned by the occupant changes it from personal property to real property. Historically mobile homes have been regulated and taxed as personal property because of their size and presumed mobility, but tax and regulation policies need to be reconsidered as the size of mobile homes increases and their mobility withers.

Trailers

During the 1920s the increasing availability of automobiles encouraged more and more American families to enjoy weekend and vacation trips on their own. Before World War I, vacations had meant trips by train or boat to conventional resort hotels, but private cars enabled people to go where they wished. Many of the areas they visited were poorly prepared to receive them, however, and families often had to camp out for lack of suitable overnight accommodations. Many carried canvas tents they could attach to one side of the car.

The tent and all the camping gear cramped the passenger space in the car, however, and setting up and taking down the tent was a time-consuming chore. Then somebody got the bright idea of packing all the gear into a wagonlike trailer that could be towed behind the car. The early trailers had sides of canvas or wood that could be raised for the night or folded down for travel, but it was only a matter of time until the wooden sides were fixed permanently in place.

The first travel trailers were little more than wooden tents on wheels, and many were homemade. It was common for early trailer manufacturers to get into the business serendipitously when they realized that there was a market for the kinds of trailers they had built for themselves. For example, Arthur G. Sherman, the president of a pharmaceutical-manufacturing company in Detroit, Michigan, is generally credited with having started the trailer-manufacturing industry almost by accident.

In 1929 Sherman planned to take his wife and five children on a camping trip to the Upper Peninsula of Michigan, but he wanted to spare himself the chore of putting up and taking down a tent each night, so he looked around for a ready-made tent he could tow behind his car. He found a wheeled box with a tent the manufacturer said could be erected in five minutes, but after sweating over it for an hour he decided that it would be easier to hire a carpenter and build his own camping unit.

Sherman's wooden box on wheels was 9 feet long and 6 feet wide, with folding upper and lower bunks and a coal-burning stove. The trap door in the rear had to be dropped to the ground to make room to cook. Nonetheless, this trailer aroused so much interest that Sherman realized he was on to

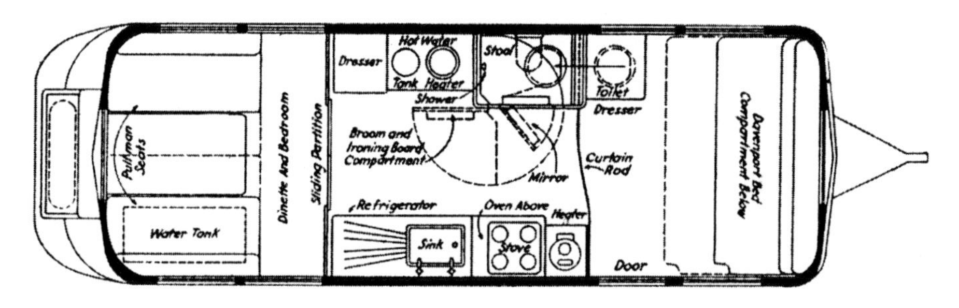

Fig. 4. Interior photograph and floor plan of a 1937-model trailer. Photograph and drawing courtesy of *Fortune*, March 1937, 108.

a good thing, and he decided to risk up to $10,000 in the trailer business. He rented a garage, hired a couple of cabinetmakers, and started building units to sell for around $300.

By 1933 business was so good that Sherman moved his operation to an abandoned candy factory in Mount Clemens, and within three months he had put an end to that city's unemployment problem. In 1936 he sold 6,000 units and grossed $3 million. *Fortune* magazine estimated that his company was the largest in an industry of some 400. The average output per factory was 250 units a year, although most "factories" were no more than side-street garages or carpenter shops that built less than half a dozen.

The trailer-manufacturing industry patterned itself after the automobile industry, and some automobile companies actually considered the possibility of getting into the trailer market, but it was too small and its low cost of entry made it unduly competitive. In a trailer factory, workers attached components and subassemblies to each unit as it moved along the assembly line on its own wheels. Some components were fabricated in shops alongside the assembly line, but others, such as bathroom and kitchen fixtures, were bought from suppliers. The rounded aluminum bodies of manufactured trailers distinguished them from boxlike homemade units, but they were still undeniably trailers.

In 1940, highway regulations varied from state to state, but they generally restricted the width of trailers to 6½–8 feet and their length to 17–21 feet. This constraint taxed the ingenuity of designers, who borrowed ideas freely from boats, from railroad Pullman cars, and later from airplanes. They moved the door from the rear to one side in order to divide the interior into a sleeping area at one end and a kitchen and living area at the other, and they tinkered with all manner of expandable models, but as long as trailers remained travel vehicles that were towed by private cars, they were constrained by highway regulations.

Trailer Parks

Trailer owners needed places to park overnight when they were traveling. In the very early days they simply stopped beside the road or in a convenient field, schoolyard, churchyard, or even cemetery, but the local people complained that they trampled sites and left too much trash. In the early 1920s many small towns on major highways decided to try to drum up business by developing municipal campgrounds. At first these campgrounds were free, but they attracted undesirables who stayed until they were chased away, and by 1924 most trailer campgrounds were charging fees. Soon they excluded tent campers, and trailer camps were called "parks" to distinguish them from lesser campgrounds.

Filling stations, grocery stores, and other businesses converted vacant lots

Fig. 5. Trailer park slum in Detroit, Michigan, in the 1930s. Photograph from *Trailer Caravan*, March 1936, 13.

out back into private pay campgrounds. A good campground had drinking water, toilets, showers, a laundry, and street lights, but many were less than optimal. Their "streets" were unlighted and unpaved, quagmires after rains, stretches of dust the rest of the time. They had privies for toilets, waste disposal was rudimentary, and extension cords carried electricity to the individual units.

Despite their deficiencies, trailer parks began to attract more and more permanent residents. The first trailers were intended for vacation trips, but almost from the start they were used as permanent residences by traveling salesmen, itinerant workers in construction and agriculture, and other mobile folk. Few new conventional houses were built during the Great Depression, and the shortage of affordable housing plus financial stress forced ever larger numbers of people to become permanent trailer residents.

Trailer parks became controversial. *Fortune* magazine derided them in 1937 as "crowded rookeries of itinerant flophouses." Critics complained that trailer residents did not pay their fair share of taxes for public services, that trailers depressed the value of adjacent properties, and that trailer people threatened the stability and morality of the community. Many municipalities escalated their license fees for trailers, put limits on the length of time they could remain in town, and passed other restrictive ordinances that discrimi-

Fig. 6. Camper trailers parked around Pleasant Lake in northeastern Indiana

nated against them. New trailer parks were permitted only beyond the municipal boundaries, although later they were grandfathered when the expanding city annexed them.

Regulatory agencies and the courts of law were bedeviled, as they still are, by the vexed question of whether trailers should be regulated and taxed as vehicles or as houses, a question that is complicated by the easily identifiable appearance of trailers and by the widespread prejudice against them. Normally trailers were appraised like automobiles, using a "blue book" that assigned them a value based on their make, model, and year of construction, and they were regulated haphazardly.

People v. Gumarsol was a highly publicized early test case. Hildred Gumarsol was a factory mechanic in Pontiac, Michigan. In the summer of 1935 he parked his travel trailer on a rented lot in the village of Orchard Lake, took off the wheels, jacked it up on blocks, and added a porch. Instead of removing it in the fall, he left it there all winter and returned the following summer. Angry neighbors brought suit, charging that he was violating a village ordinance by living in a dwelling unit with less than 400 square feet of floor space. Gumarsol retorted that his trailer was not a dwelling, because it was licensed as an automobile accessory. Justice of the Peace Arthur Green ruled that the trailer was a dwelling but fined Gumarsol only $1 plus $3.10 costs, "because we knew that this was an important test case."

Gumarsol's was not the only trailer in the village, and Justice Green was obviously concerned lest Orchard Lake become a "trailer shantytown," but other considerations were also involved. "The trailer people roll in their trailers and proceed to enjoy all of the privileges of the lake without paying taxes," snorted Police Chief Clarence Carson, "and they aren't too discreet about getting into bathing suits, either." Others fretted that children living in one-room trailers acquired a precocious knowledge of sex.

Trailer manufacturers were not happy with the outcome of the Gumarsol case, because they wanted trailers classified as vehicles rather than as housing units. They saw their hope for the future in commercial use, with trailers fitted out as traveling showrooms, mobile medical clinics, libraries, and the like. In 1937 *Fortune* magazine said that trailer builders were disturbed by the very idea that trailers could be used as permanent housing, a notion they considered nonsensical. World War II changed all that in a hurry.

The War Years

"Significant advances in the use and design of mobile homes," wrote Allan Wallis, historian of the mobile home industry, "have occurred primarily in periods of unmet housing demand," and it was assuredly the housing crisis of World War II that transformed the popular perception of trailers from vehicles for vacation travel to mobile homes for year-round residence. People who had used their trailers only for vacations paid outrageous rental fees to park and live in them in someone's backyard near the war plant, and their pride in contributing to the war effort overcame the stigma of living in a trailer.

Hordes of workers flocked to areas where defense-related activities grew explosively during the war, and many of these areas suffered acute housing shortages. Richard Foster reported that in the San Francisco Bay area in 1943, for example, "small retail stores were converted to living quarters; trailers, tents, tin houses, cardboard shacks, barns, garages, automobiles, theaters, chicken coops, and open fields served as residences, and workers sometimes had to sleep in shifts in the same bed."

Under these circumstances travel trailers looked almost palatial, and every trailer in sight was drafted into service as a year-round residence. One-room travel trailers had to be in parks with washing and toilet facilities, because they did not have showers or even toilets. There were never enough parks, even though some municipalities relaxed their restrictions against them, at least for the duration of the war, and life in a wartime trailer park could be a harrowing experience.

In 1940 various government agencies began buying trailers for temporary "stopgap" housing until conventional housing could be built in areas impacted by heavy influxes of war workers. Trailer manufacturers had to grit their teeth and agree that trailers were houses in order to obtain scarce or critical building materials that were subject to wartime rationing, but they boasted that they employed semiskilled and elderly workers who could not otherwise contribute to the war effort.

Four-fifths of the standard GI trailers were 8-by-22-foot units, each equipped with a stove, a refrigerator, and two studio couches that opened into double beds, but they lacked running water. The rest were expandable units with two wings that could be pushed out to provide sleeping space for six people, and they had their own bathrooms.

In 1942 all the civilian housing activities of the federal government were consolidated into the National Housing Agency (NHA), which at one time had cognizance of more than 35,000 trailers. The administrators of the NHA held strongly traditional ideas about what constituted proper housing. They disdained trailers as substandard and stopped buying them in 1943. After the war the NHA gave 13,000 surplus wartime trailers to colleges and universities for married-student housing, which presumably did not endow these students with a more favorable disposition toward the advantages of trailer living.

Few parts of the United States were as heavily impacted by World War II as the San Francisco Bay area, where shipyard employment alone increased from 10,000 in 1940 to 200,000 in 1943. Workers poured in from all parts of the country and overwhelmed the local supply of housing. More new housing units were completed in the Bay area during the war than in any other metropolitan area in the country, but housing was always tight.

The war's influence on the geography of trailer parks is particularly clear in the Richmond, California, area, which had 4 large shipyards and more than 50 other defense industries (Fig. 8). The prewar parks were on major highways and served travelers who stayed only a few days. These parks filled immediately with war workers, and they were supplemented by 21 new parks near the shipyards and other industrial employers. The new parks served workers who expected to remain in the area for the duration of the war.

Fig. 7. The Aggie Villa Mobile Home Park in Goodwell, Oklahoma, rents homes to students at Panhandle State University

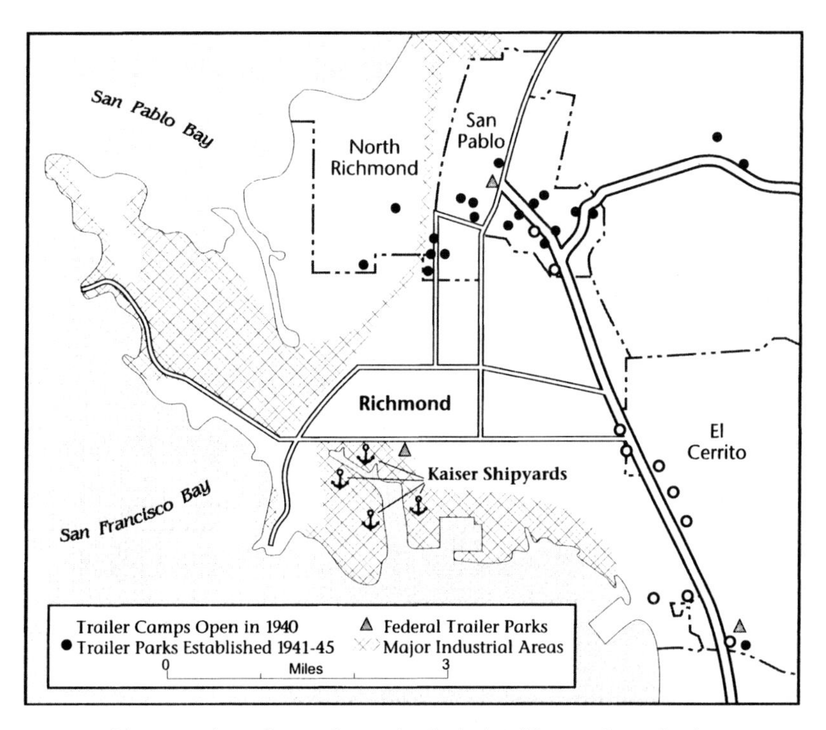

Fig. 8. World War II trailer parks in and around Richmond, California. After Richard H. Foster Jr., "Wartime Trailer Housing in the San Francisco Bay Area," *Geographical Review* 70, no. 3 (1980): 284.

Most of the new trailer parks were in San Pablo, an unincorporated town in Contra Costa County, where trailer parks were permitted, regulations were leniently enforced, and semirural properties of five acres or so were available. Richmond, in contrast, had no private trailer parks, despite the fact that city ordinance permitted them, because the city council routinely rejected applications for new parks on the grounds that they were undesirable.

The new trailer parks were small but crowded. Owners tried to pack as many trailers onto their land as possible. Few parks had as many as 50 trailers, but some crammed as many as 40 onto a single acre. The parks were so crowded that people got on each other's nerves, both inside and outside their trailers. Whites from the South were offended that black families were allowed to live in the same park. Unsupervised children running free with their slingshots were a problem when both parents were working full-time. Neighbors complained that some park residents were undesirable types, and they fretted about the continuous turnover of residents. Residents and outsiders alike were angered when parks were not well maintained.

Although the Richmond City Council did not allow private trailer parks, it could not stop the federal Public Housing Agency from establishing a public trailer park practically across the street from the shipyard. The federal trailer parks were designed as models, with better facilities, community buildings and playgrounds, organized recreational and social activities, and the gamut of low-order central-place functions. They were closed as soon as the war ended, and the land was returned to the owner from whom it had been leased for the duration.

Even though the government closed down its trailer parks, gave away its trailers, and got out of the trailer business as soon as hostilities had ended, World War II legitimized the use of trailers for regular residences. It helped that trailers built after the war had indoor plumbing, but they still had to be hooked up to utilities. New and larger trailer parks were developed to serve them, but zoning regulations in many polities restricted these parks to commercial or industrial areas along railroad lines or near factories, car dealerships, junkyards, or even sewage farms. Relegating trailer parks to unattractive and undesirable sites did nothing to improve their unfavorable image. They were permitted on flood plains on the assumption, invalid even then, that the units could easily be towed away if ever they were threatened by a flood.

Fig. 9. A mobile home that houses Hispanic workers on a new hog farm north of Springfield, Colorado

The postwar housing shortage forced many people to live in trailers, which became the residence of choice for itinerant oil field workers and migrant farmworkers. Whole crews of construction workers moved in trailers from job to job on pipelines, mines, dams, the interstate highway system, and Atomic Energy Commission facilities. Nearly 10,000 trailers were clustered near the AEC facility outside Augusta, Georgia, at the peak of construction, and many military bases are still festooned with trailer parks.

The government has continued to use trailers as quick and easy emergency housing for victims of floods, tornadoes, hurricanes, and other disasters. The Federal Emergency Management Agency housed 3,500 displaced families in 12 trailer parks after Hurricane Andrew devastated the area south of Miami in 1992, and the last of these parks was not closed down until two and a half years later. Trailers have also been used for business purposes or for mobile offices on construction sites. Mobile classrooms are almost standard additions to schools to alleviate classroom overcrowding, and California State University, Northridge, set up offices and classrooms in an armada of trailers after an earthquake shattered the campus in January 1994. In 1996 the police department in Fort Lee, New Jersey, was housed in four trailers jammed into a parking lot because plans to build a new police station had become bogged down in small-town politics.

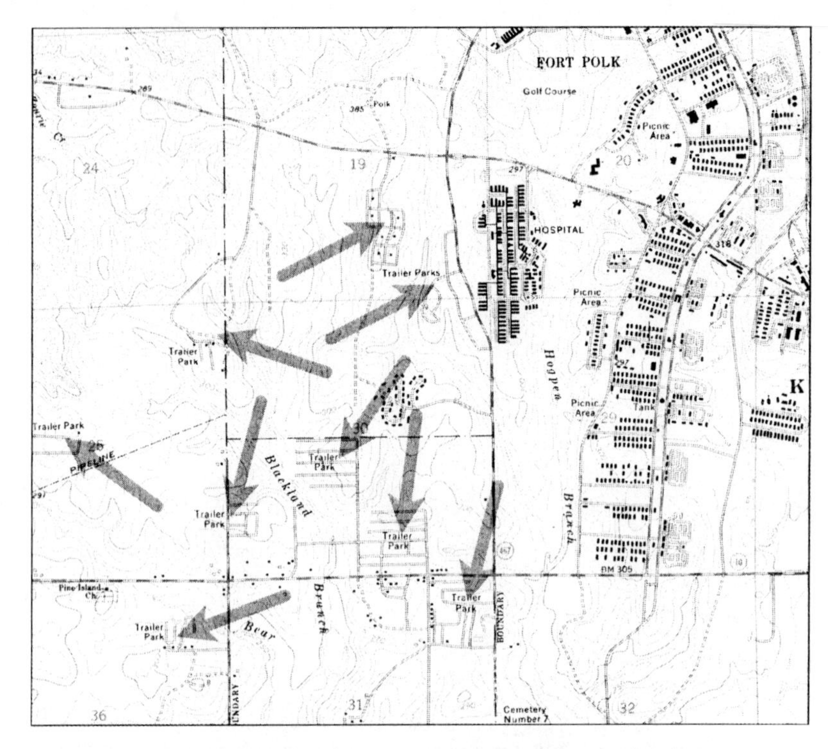

Fig. 10. Excerpt from the Fort Polk, Louisiana, 1:24,000 topographic quadrangle (U.S. Geological Survey), showing trailer parks outside the main gate of the Fort Polk army base

Fig. 11. This mobile home has been pressed into service as the post office in Itmann, West Virginia, a coal-mining camp

Mobile Homes

The trailer industry was of two minds after World War II. Trailer manufacturers managed to keep themselves in business by making year-round housing units for the special subset of the population that needed mobile homes, but they were really waiting patiently for the return of their "normal" market for vehicles for recreational travel. When a market for travel vehicles actually did begin to develop, however, it turned out that customers wanted self-propelled recreational vehicles (RVs) rather than trailers that had to be towed, and the trailer makers had to accept the fact that their future lay in making houses rather than vehicles. (In 1963, in fact, the industry actually did split into an RV trade and lobby group and a mobile home group.)

A market survey in 1948 found that most trailer people were construction workers or military personnel, but a similar survey in 1959 revealed that trailers had become preponderantly starter homes for newly formed families with young children. These young families wanted units that were more like houses and less like tents, boats, trains, or planes. They were annoyed by collapsible and folding furniture and fixtures, no matter how clever these devices might be, and they wanted conventional fittings that would make the unit seem as much as possible like a house.

The people who lived in trailers year-round were more concerned with livability than with mobility. They wanted units that were larger, more spacious, and less cramped than travel trailers, but the size of trailers in the early 1950s was pretty much determined by state laws governing the size of vehicles that could be towed. These laws varied from state to state, but most states only permitted units that were no more than 8 feet wide, 12½ feet high, and 35 feet long.

Trailer manufacturers were challenged to make this space as livable as possible. Some tried to enlarge it by building expandable sections that could be folded out or telescoping roofs that popped up, but these ideas never really caught on, because the joints had the unpleasant habit of leaking water and cold air. Many manufacturers thought the answer lay in longer models, despite their reduced maneuverability, and they lobbied state legislatures for more permissive laws.

Elmer Frey, who started building trailers in Marshfield, Wisconsin, in 1947, had the revolutionary idea of making units that were wider rather than

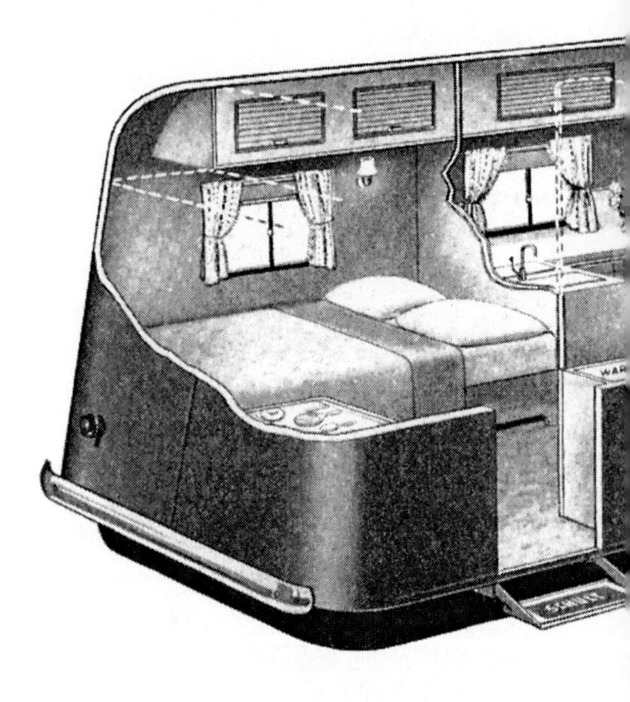

longer, and in 1954 he introduced a model that was 10 feet wide. He had to use a railroad flatcar to ship it to the all-important Florida Mobile Home Exposition in Sarasota, because it would have been impractical to try to get all the permits he would have needed to have it towed by highway across six or seven states.

Frey's new 10-wide was an immediate success, because the additional width conferred greater privacy. The middle bedroom and bathroom in a conventional 8-wide had to be corridors to the back bedroom, but a 10-wide had sliding doors that separated them from a passageway down the side. A 10-wide also had more storage space, and the refrigerator, washing machine, and furnace were moved out of the kitchen to leave more room for the dinette.

Frey argued vehemently that a 10-wide was not a towable trailer but a mobile home, and that it should be given the same oversized-load permits as tractors, combines, boats, and construction shacks. By 1957 most states had agreed to allow 10-wides to be towed on their highways during daylight hours, but they placed restrictions on the towing vehicle that eliminated the

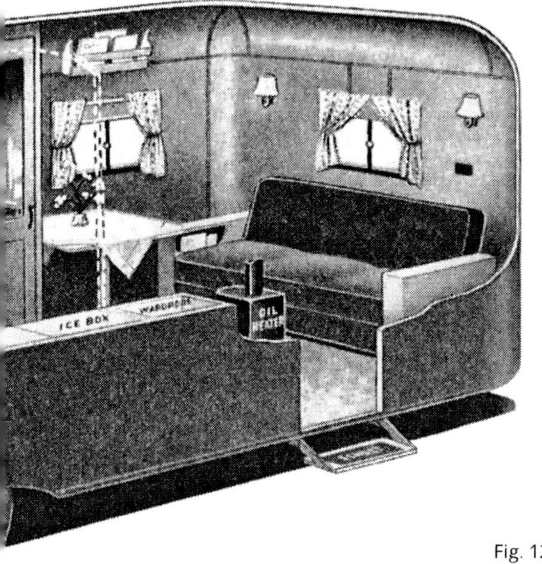

Fig. 12. Cutaway drawing of a 1943-model trailer. From *Trailer Travel*, December 1943, back cover.

Fig. 13. Interior view of the kitchen and dining area of a 1937-model trailer. From *Trailer Caravan*, March–April 1937, back cover.

possibility of using the family car. In essence, the 10-wide could be towed to the site where it was going to be put, but it was no longer a travel vehicle. It was a mobile home that was going to be immobilized on a particular site.

The die had been cast. By 1960 the 10-wide had become standard, and 12-wides had been introduced. Henceforth the mobile home–manufacturing industry acknowledged that it was making dwelling units rather than vehicles. It had long been trying to improve the negative image of its products. As early as 1953, in fact, the Trailer Coach Manufacturers Association had changed its name to the Mobile Home Manufacturers Association, but the name change was slow to catch on colloquially, and skeptics perceived it as mere name inflation.

Manufacturers did their best to make the interiors of mobile homes as much as possible like traditional houses. In a trailer the kitchen and bathroom had been close together in the middle of the unit, to save on plumbing, but in a mobile home the kitchen was at the front, where it was possible to watch the street for visitors and keep an eye on children. The doors and windows of mobiles were more conventional, with draperies softening the harsh and shiny surfaces of metal, linoleum, and plywood.

The exterior was more of a problem. The shiny rectangular metal box was hard to disguise, but many owners expressed their individuality with additions and changes that made their units more livable and more in line with suburban standards of taste. They built concrete walkways and steps, and added porches and sundecks. They built sheds for storage, they stretched awnings over the concrete parking slabs to create carports, and in time they screened in the carports to make outdoor sitting rooms. They planted shrubs to screen the hitch and propane bottle in front, and they hid the undercarriage with sheet metal or hardboard skirting. Some additions were so elaborate that they made the original mobile home look almost like an afterthought.

By 1969 the industry was producing units that were 14 feet wide, with a greater variety of architectural designs and materials and more amenities. Manufacturers offered standard models of various sizes and layouts, distinguished from one another by their fittings and by the number and luxuriousness of their bedrooms and bathrooms. Different models might feature walk-in closets and dressing areas, kitchens with island work centers, and cathedral rather than flat ceilings.

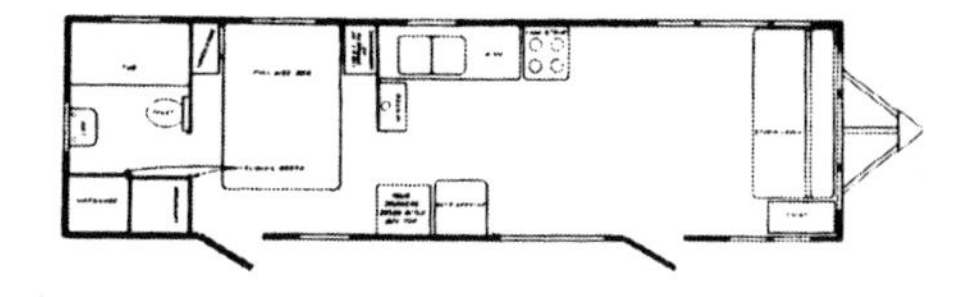

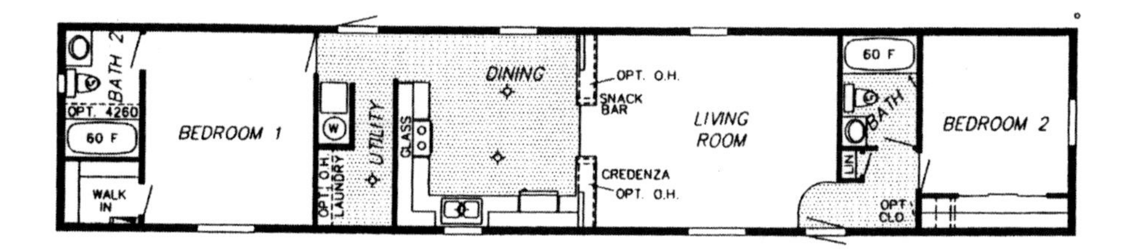

Fig. 14. Floor plans of three generations of mobile homes, all drawn at the same scale. Courtesy of Wick Building Systems, Inc.

By 1969 the industry was also producing multisectionals, or double-wides, as they are known colloquially. A multisectional consists of two or more separate sections that are towed to the site individually and assembled on the spot. Once assembled, a multisectional is seldom taken apart and moved. Many multisectionals have masonite, wood, or vinyl rather than aluminum siding, and they have pitched roofs with shingles. To the inexperienced eye today's multisectional is virtually indistinguishable from a conventional site-built house.

We have not included any photographs of the interiors of modern mobile homes in this book because they look just like the interiors of conventional site-built houses.

Fig. 15. Half of a double-wide mobile home being towed to its site

Fig. 16. Two halves of a double-wide mobile home waiting to be assembled

Fig. 17. A single-wide and a double-wide mobile home on the same site near Gainesville, Florida

The HUD Code

The 1970s were heady days for the mobile home industry, because the federal government had begun to recognize mobile homes as a legitimate form of permanent housing. Inadequate low-income housing was one of the causes of the terrifying urban riots in the summers of 1965–67. The government considered a variety of programs, including greater use of mobile homes, to alleviate the shortage, but the only result was a low-interest loan program. This program did not specifically suggest mobile homes, but their sales skyrocketed because no other form of housing was as affordable as mobile homes, even though they had no government subsidy.

It was clear that the country needed mobile homes, but many people were concerned about their safety and quality. They were uniquely susceptible to damage by fire and by high winds, and some were downright flimsy. The great majority of manufacturers were reputable, but a few used the cheapest and shoddiest materials they could find, and their units were assembled by unskilled workmen.

In 1974 Congress responded to these concerns by passing the Mobile Home Construction and Safety Standards Act, which authorized the Department of Housing and Urban Development to establish and enforce a national building code that would ensure that mobile homes satisfied minimal standards of safety and durability. The HUD Code, which was promulgated in 1976, covers the design and construction of mobile homes and their fire-safety, plumbing, heating, and electrical systems. It was preemptive, in that it took precedence over state and local building codes.

The HUD Code has been a useful marketing tool for the mobile home industry, because a unit that carries the red and silver HUD label certifying compliance can be sold anywhere in the country. Federal certification has also made mobile homes more attractive to financial institutions, which has facilitated financing at rates closer to those available to conventional home-buyers and has helped to open secondary mortgage markets.

Critics agree that the HUD Code has greatly improved the safety of mobile homes, but they complain that HUD oversight of design and production has been weak, and its inspection system is seriously flawed. Instead of hiring federal inspectors, HUD merely approves the qualifications of local inspectors who are actually hired and paid by the manufacturer. Some inspectors

Fig. 18. The steel chassis is all that remains of a mobile home that was blown away by a windstorm

Fig. 19. Mobile home battered by a windstorm

are not particularly well qualified or well trained, and there is a major potential for conflict of interest when inspectors are on the payrolls of manufacturers whose designs and production they are supposed to inspect.

Consumer advocates have been critical of the glacial pace at which HUD has moved to correct deficiencies in the code and to update it to keep up with innovations in building technology, such as improvements in energy efficiency. They also claim that HUD has failed to protect consumers adequately against unfair practices. For example, most mobile home warranties exclude damages resulting from transportation and installation, although that is precisely when damages occur. The manufacturer blames the dealer and the dealer blames the manufacturer, leaving the poor consumer with no recourse but to take them both to court, which can be even more expensive than repairing the damage.

On the other side, manufacturers claim that the HUD Code is too rigid, restrictive, and stifling. It requires all units to have a permanent steel chassis, which the manufacturers consider unnecessary, but conventional builders say that changing this rule would put them at an unfair disadvantage. The manufacturers also complain that the costs of the bureaucratic paperwork associated with the code have driven many small companies out of business or forced them to sell out to larger competitors. The shakeout of small companies has always been a fact of life in the mobile home–manufacturing business, but in recent years a few large companies have become increasingly dominant.

The industry has identified a comfortable niche in the American housing market by concentrating on upscale models that are ever more conventional and houselike. The lean and hungry look of the trailer and mobile home has been replaced by the sleek, handsome, prosperous aura of manufactured houses, which are built in factories and towed to their sites but are otherwise indistinguishable from conventional site-built houses and give scant hint of their genealogy. The champion thus far is a multisectional in Florida that has an indoor swimming pool and cost $300,000, but it is sure to be eclipsed eventually.

New Models

Mobile home dealers used to joke that most of their customers were "either newly wed or nearly dead," but the industry has been inching upward into the middle-income market and away from starter and retirement homes for low-income people. In 1994 half of all mobile home owners were aged between 30 and 49, half had household incomes of $20,000 to $40,000, and half of all new units were houselike multisectionals, or double-wides.

The industry signaled this change in 1975, when the Mobile Home Manufacturers Association changed its name to the Manufactured Housing Institute, and the 1980 Housing Act decreed that "the term mobile home be changed to manufactured housing in all federal law and literature." The new name has not caught on particularly well outside the industry, and despite its best efforts many people and even some federal agencies still persist in using the terms "mobile home" and "trailer."

Allan Wallis has concluded sadly that the industry has become conservative and complacent. In its quest for legitimacy and respectability, it has become more conventional and less flexible, and it is no longer innovative at the lower end of the market. It has moved upscale and has abdicated its traditional social role of providing affordable housing for low-income households.

Wallis advocated greater attention to small one-bedroom units with 320 to 400 square feet of floor space, which are too small for the HUD Code but too large to be classified as recreational vehicles. "Park" models are designed for small lots in older parks and for easy maintenance, which especially appeals to older and seasonal occupants, but they are fairly expensive because they are intended for the affluent second-home market.

Elder Cottage Housing Opportunity (ECHO) models, or "granny flats," are designed for older persons and those with disabilities who need supportive services but are not yet ready for nursing homes. These units can be installed next to the single-family homes of family or friends who can give the occupants the support they need to live independently, and the units can be towed away when they are no longer necessary.

ECHO units usually require a zoning variance, which is often hard to get, but the American Association of Retired Persons (AARP) has developed a persuasive legal argument for permitting ECHO units even where zoning and covenants restrict the land to single-family residences. The AARP also suggests that a person who sold a mortgage-free home for $50,000 and bought an ECHO unit for $28,000 could enjoy additional income of $146 a month if the difference of $22,000 were invested at 8 percent.

Siting Mobile Homes

People who buy new mobile homes must have some place to put them. The customary advice is to secure the site before you buy a unit, because not all units will fit on all sites. Nobody really knows for sure, but the best guess seems to be that slightly less than half of all mobile homes are clustered in parks (some of them land-lease communities where residents own their units but rent their sites, others freehold communities where residents own

Fig. 20. A double-wide mobile home resting precariously on cinder blocks

Fig. 21. Upscale mobile home park north of Englewood, Florida

both the unit and the site), and slightly more than half are single-sited on separate parcels of land, most on the owner's private property, but some on property that belongs to someone else, usually a relative or friend.

Single-sited mobile homes often run afoul of local regulations designed to exclude them by imposing restrictions on features not covered by the HUD Code, such as minimum length/width ratios and floor area, foundations, exterior siding, window size and style, and roof pitch and style. Mobile homes can easily match conventional roof pitch of 4/12 (one foot of rise for every three horizontal feet), but steeper roof pitch can make units more difficult to transport because of height restrictions on highways. About half the states have laws prohibiting zoning that discriminates against mobile homes, but locally imposed restrictions must often be challenged by expensive court cases that most individual mobile home owners cannot afford.

Recently an experimental program in six cities has single-sited mobile homes as urban infill on the kinds of small narrow lots that are scattered through central-city areas. It is easy to roll in a new mobile home and set it up on such a site. Some are on vacant lots, and others have replaced dilapidated older houses. The new units are often less expensive and more livable than the substandard and poorly maintained older houses they replace.

Each mobile home park has its own distinctive character. Parks range from the kinds of places that have given them a bad name to luxurious estates far too palatial to be called mere "parks," though thus far no one has been able to come up with a better generic name for them. Most of the newer and more luxurious parks are in retirement areas in Florida, California, and Arizona, and they are dominated by multisectional units.

Utilitarian older parks dot the fringes of every city and town, and even many villages, because land on the fringe is available, it is cheaper, and it is subject to fewer restrictive zoning regulations. In most areas, however, park space is in short supply because some older parks have been engulfed by the city and converted to more lucrative uses, and farmland preservation programs have hindered the development of new parks.

The first trailer parks were intended for transients and short-term residents. Easy entrance and exit was a paramount consideration. Their lots were small and narrow, just large enough to hold a trailer. They were laid out at right angles to long straight streets, and from your living room window you could look through the living room windows of all the units on the street. The trailer hitch constituted the front yard.

When people took up permanent residence, they wanted parks that looked more like suburbs and mobiles that looked more like houses. Parks for mobile homes had to have larger lots as the units got larger, and in many parks the lots were laid out diagonally to the street, in herringbone fashion. Curving the streets even gently helped to make parks look less monotonous, and lots were clustered around culs-de-sac when it became apparent that units were no longer likely to be moved. Some parks require that the hitch be camouflaged, the undercarriage be hidden by skirting, and the lot be landscaped. Some even require site-built additions, such as carports and covered patios.

Fig. 22. Utilitarian mobile home park near Gainesville, Florida, with concrete pads for more homes in the foreground

Fig. 23. Mobile home park with curving streets and culs-de-sac near Lansing, Michigan

Single-sited mobile homes and mobile home parks both have continued to evolve to serve an ever wider spectrum of income and ethnic groups. Single-wides and multisectionals are commonplace in all corners of the country except for inner cities, and even there some mobile homes are being integrated into the housing stock. The owners of mobile homes treat them just as they would treat site-built houses. They paint, build, loft, plant, raise, tear down, add on, and make other changes to tailor their homes to their needs. Most mobile home owners are settled in for the long haul.

PART TWO

SINGLE-SITEDS

Where Are They?

Statistics on mobile homes in the United States are probably more accurate and reliable than statistics on the incidence of adultery, but many are little better than educated guesses. For example, take the estimate (which we ourselves have invoked) that 95 percent of mobile homes are never moved again after they are first sited. Nobody can prove that it is either right or wrong, because nobody really knows.

Information about the number and distribution of mobile homes in the United States is available from private organizations, from the U.S. Bureau of the Census, and from the U.S. Geological Survey. Most of this information is collected by private organizations that have neither the data-collecting know-how nor the coercive authority of federal agencies. These organizations produce the best information available, and we must be grateful for their efforts, but some of their estimates, unfortunately, must be taken with a rather large grain of salt.

For example, the Manufactured Housing Institute collects and publishes useful data on the number of mobile homes built each year. These numbers are a good measure of changes over time, but they are not available for geographic units smaller than major regions of the country. The Foremost Insurance Company of Grand Rapids, Michigan, estimated that 37 percent of all mobile homes in the United States in 1990 were in rented spaces in parks, 6 percent were in owned spaces in parks, 39 percent were single-sited on the owner's private property, and 17 percent were single-sited on rented property, but these estimates likewise were applicable only to the entire nation, and local variations were legion.

In 1990 the U.S. Bureau of the Census treated mobile homes as an awkward afterthought and relegated them to a residual category. In published reports for small areas it lumps "mobile homes or trailers" with "other" in a category that also includes houseboats, railroad cars, camper vans, recreational vehicles, permanently inhabited tents, and other nontraditional forms of housing.

The numbers in the "other" category might also have been inflated by an awkwardly phrased census question. At each housing structure the census enumerator asked the occupant, "Which best describes this building?" and offered a list of possible responses that included "a mobile home or trailer" and "a one-family house detached to *[sic]* one or more houses" as well as "other." The occupant of a single-family detached house might have opted for "other" if confused by that second choice. The 1990 Census of Housing, for example, reports that the village of Ephraim in Door County, Wisconsin, had 75 mobile homes, trailers, and other housing units, even though the village has a strict ordinance against mobile homes.

Furthermore, census data on mobile homes and trailers, unlike most other census data, probably are not comparable from census to census, especially for small areas. We compiled maps of change over time at the county level, but we suspect that these maps may be more misleading than helpful. Despite the flaws of census data, however, these are the only data available for mapping the distribution of mobile homes in most of the United States. We must use these data, but we must be sensitive to their quirks, and we must interpret our maps with circumspection.

The topographic quadrangles published by the U.S. Geological Survey show the specific locations of selected mobile homes and parks (Figs. 10, 68, and 72). In the late 1950s the survey developed mapping doctrines that have not changed much since then. Isolated trailers or mobile homes on permanent foundations may be shown as buildings. Manufactured homes that look like permanent buildings are mapped like permanent buildings. Mobile home parks are identified if they have at least one mapped road. They were labeled "trailer park" before 1994, "mobile home park" subsequently. Individual mobile homes are not shown because of their small size, close proximity, and potential for being moved, but installations, such as permanent buildings or swimming pools, are symbolized. Parks in urban areas were labeled and not tinted with an urban color before 1994, but since then they have been tinted and not labeled.

We need a pair of maps to show the importance of mobile homes in the United States in 1990 (Figs. 24 and 25). The map of mobile homes per square mile emphasizes their total numbers, and the map of mobile homes as a percentage of all housing units emphasizes their relative importance. For example, in 1990 Orange County, California, had 31,843 mobile homes, and Eureka County, Nevada, had only 502, but 60 percent of the homes in Eureka County were mobile homes, in contrast to fewer than 4 percent in Orange County. Orange County had 323 mobile homes on every eight square miles, but Eureka County had fewer than one. Both measures are equally important.

In 1990 the densest concentrations of mobile homes were in resort and retirement areas, near metropolitan areas, in the manufacturing belt of the Southeast, and in the coal fields of Appalachia. Ten percent of the nation's mobile homes were in Florida, and mobile homes were concentrated in other resort and retirement areas, especially in those close to water bodies, whether along the Atlantic and Gulf coasts, on the natural lakes of the Upper Lake States, or on manmade impoundments elsewhere.

To a certain degree the distribution of mobile homes reflects the overall distribution of population, because the greatest numbers are in areas where the greatest numbers of people need them. One-quarter of the nation's mobile homes in 1990 were in metropolitan areas. Virtually every major metropolitan area, as well as some that are not so major, had a significant cluster, and many college and university towns were also singularly well supplied. Often the greatest numbers were not in central cities or even in central counties, which had relatively few, but in perimetropolitan counties with permissive zoning regulations. St. Louis illustrates this "hole in the doughnut" effect especially well (Fig. 26).

Mobile homes were not uniformly distributed in the perimetropolitan periphery, however, because some counties seem not to object to them, whereas the "better" counties seem dead set against them. For example, in 1990 affluent Westchester County, New York, had only 192 mobile homes, but Rockland County, on the other side of the Hudson River and only two-fifths as large, had 1,148. In Illinois, Kendall County, suburban to Chicago, had only 43 mobile homes, but Lake County, only slightly larger, had 3,446.

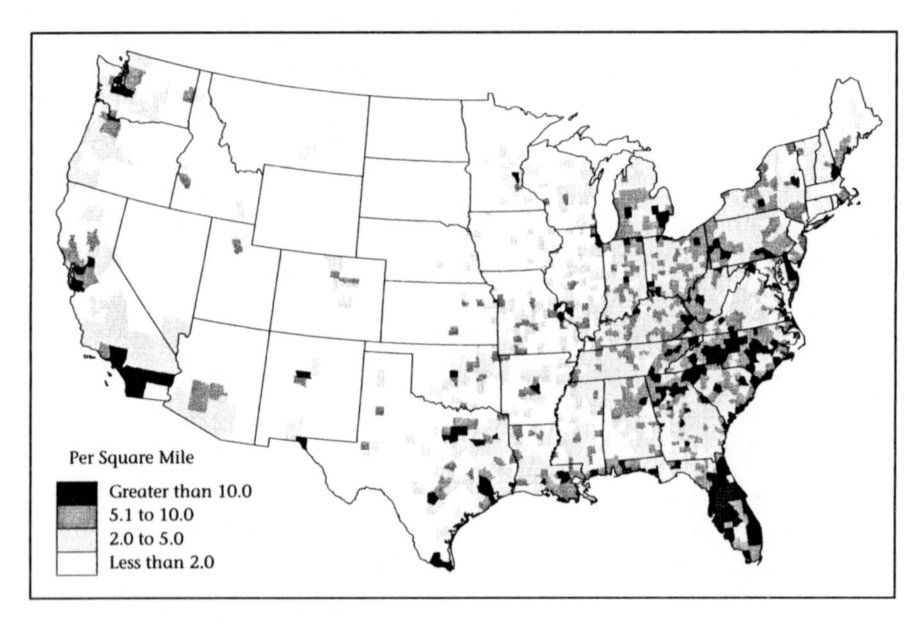

Fig. 24. Mobile homes per square mile in the United States, 1990

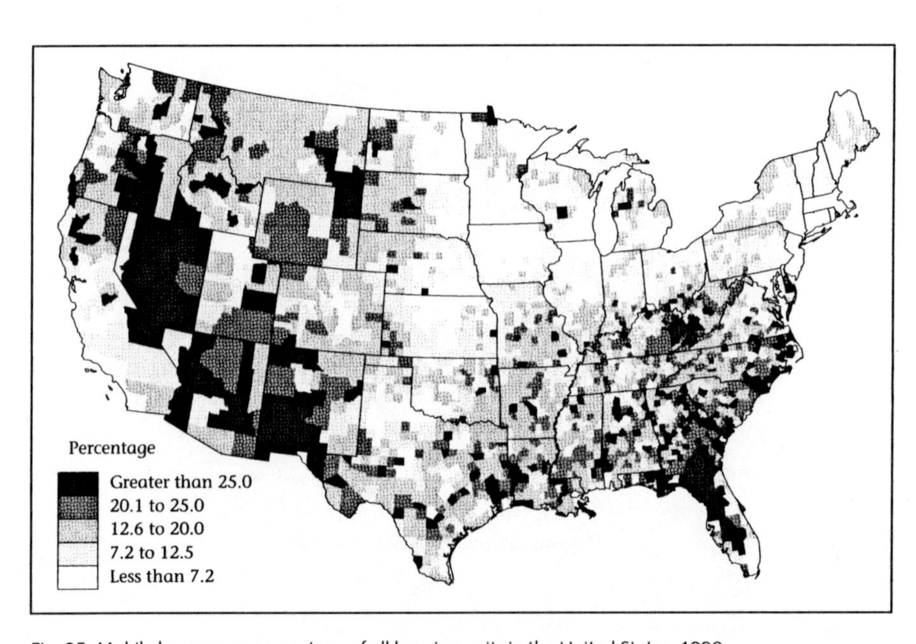

Fig. 25. Mobile homes as a percentage of all housing units in the United States, 1990

The manufacturing belt of the Southeast, which includes both the Piedmont of the Carolinas and Georgia and the Great Valley of northeastern Tennessee and southwestern Virginia, had 7 percent of the nation's mobile homes in 1990. This area, which we call "Spersopolis" (the dispersed city), has developed a distinctive new settlement pattern that might well be the model for the populist American city.

Although metropolitan areas had one-quarter of the nation's mobile homes in 1990, these homes comprised less than 3 percent of their total housing stock. The percentage of mobile homes increased as the size of place decreased. They comprised more than 10 percent of the housing stock in incorporated places of fewer than 2,500 people, and 20 percent of the housing stock in rural areas, which had 4.2 million mobile homes, 56 percent of the national total.

In the West and South, mobile homes seem to be accepted with greater equanimity than in the North wherever low-cost housing is needed, especially when it is needed quickly. They form a significant share of the housing stock of remote, sparsely populated, low-income rural areas, although high percentages can be deceptive in the West, where counties are large and people are scarce. Many low-income and poverty areas have high percentages of mobile homes, but not all counties with high percentages of mobile homes are associated with low income, because some people use them as affordable homes in recreation, resort, and retirement areas.

Census data do not confirm widespread myths about racial attitudes,

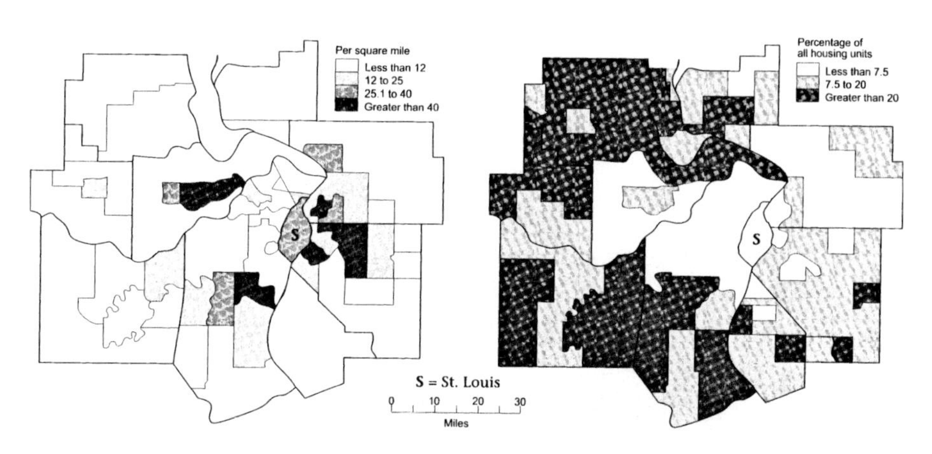

Fig. 26. Mobile homes in the greater St. Louis area, 1990

Fig. 27. Mobile home used as a farmhouse on a dairy farm south of Osseo, Wisconsin

Fig. 28. Mobile homes interspersed with conventional houses in Brookings, South Dakota

whether positive or negative, toward mobile homes, and any associations of blacks, Hispanics, or American Indians with mobile homes can probably be explained better by income levels than by race.

In the West virtually every town has at least one mobile home park, and mobile homes are scattered through the town on single lots intermingled with conventional houses. Almost any new economic development in the region requires instant housing, and new industries, such as food-processing plants, have had to import and house the workers they need. Immigrant workers, particularly Hispanics and Asians, have encountered hostility from long-term residents, and companies have had to develop mobile home parks to house them (Fig. 9). Of course, company housing also facilitates control of the workers.

Mobile homes are ubiquitous in the rural areas of the West and South, with especially high percentages in or near some but not all mining areas (such as the southern Appalachian coal field or the Gulf Coast oil patch), major construction sites, military bases, and resort areas. One might expect that mobile homes near construction sites and in mining areas would be temporary, but the number of mobile homes declined in only a handful of counties in the entire United States between 1970 and 1990, and in many counties the number increased handsomely, which indicates that they have become a permanent part of the nation's housing stock in rural areas.

The Lexington Hexagon

At the local level the distribution of mobile homes and mobile home parks is idiosyncratic. At the national level we can hazard some reasonably good explanations of why mobile homes are where they are, and even at the local level it is usually fairly easy to "explain" the location of any specific mobile home or park, but generalizing about their distribution in any particular area is surprisingly difficult.

Thomas P. Field seems to have been the first scholar who explored and tried to explain the distribution of mobile homes and mobile home parks in a particular area. He compiled a map of mobile homes and parks in a hexagonal area around Lexington, Kentucky, in 1970, and made a valiant effort to specify a model that would "explain" their distribution (Fig. 29).

In 1970 the Lexington hexagon had 3,650 mobile homes, which comprised 4 percent of all dwelling units. Ten percent were single-sited, and another 10 percent were in small, informal clusters of 15 or fewer units on horse farms, at recreational sites, and near factories. One-quarter were in small parks of 18 to 68 units, and slightly more than half were in large parks of 93 to 350 units.

The average park had seven units per acre. Some parks hardly had space to walk between the units, but others had as much space as small suburban lots. The small parks were generally the oldest and of the poorest quality. Lot rents ran from $27.50 to $45.00 a month, and it appeared that tenants got what they paid for in terms of park facilities, quality, maintenance, and accessibility.

Fig. 29. Mobile homes and mobile home parks in the area around Lexington, Kentucky, 1970. After Thomas P. Field, *Mobile Homes of the Kentucky and Lexington Hexagon: A Study in Areal Distribution,* Kentucky Study Series no. 5 (Lexington: University of Kentucky Department of Geography, in cooperation with the Fayette County Geographical Society, 1972), between pp. 28 and 29.

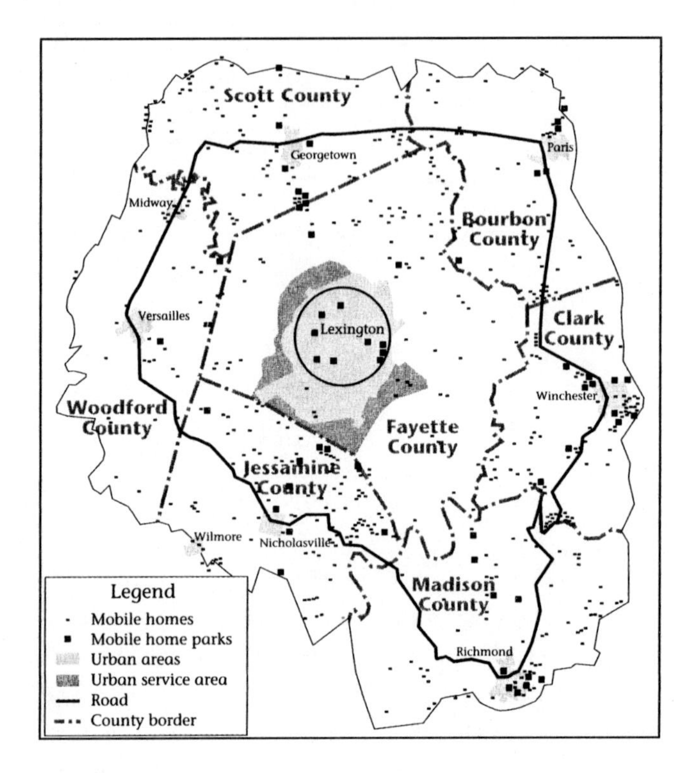

In 1970 the mobile homes of the Lexington hexagon were dwellings that people of modest income could afford. They were predominantly occupied by low-income blue-collar people, but they also provided affordable housing for married students at the University of Kentucky. They were indicators of social isolation, but not poverty. "In the public mind," said Field, "trailer people had about the same social status as gypsies." They were an unorganized and unrepresented minority group, with few ties to the general community and little political voice.

The distribution of mobile homes within the Lexington hexagon in 1970 was influenced by land values, distance from Lexington, local zoning ordinances, the regulations of the state Water Pollution Control Commission, and other factors. The densest concentration was south of Lexington in Jessamine County, which had the most permissive ordinances, but units seem to have been plunked down wherever they could find a niche that would accept them, and Field concluded that mobile home parks were simply "happenings" rather than planned developments.

Field made a valiant effort to formulate a theoretical model that would "explain" the distribution of the mobile homes and parks of the Lexington hexagon. He considered a large number of possible explanatory variables but was forced to conclude that the location of mobile homes and parks was controlled largely by regulation and by chance; the location of individual homes and parks, he said, was an "accident," the conjunction of an owner or developer and land that was available and affordable.

Field's quest for a model that would explain the distribution of mobile homes and parks in a particular area eventually proved fruitless, but no one subsequently has been able to do any better. It seems safe to conclude that the distribution of mobile homes and parks in any particular local area is random and idiosyncratic.

Northern New Mexico

Field's study is unusual in that most discussions of mobile homes in specific areas have been incidental to studies that have focused on other subjects. Much of what we know about mobile homes has had to be cobbled together from bits and pieces gleaned from a variety of separate sources from widely scattered areas, or is based on our own fieldwork.

For example, Alvar Carlson and J. B. Jackson both have written about mobile homes in northern New Mexico. Carlson said that mobile homes began replacing traditional adobe houses in the 1960s because young Spanish-American families wanted indoor plumbing, air conditioning, and other modern conveniences. Parents allowed their newly wed children to place new mobile homes near their old adobe houses, and they kept them even after the children had left so that the young people would have a place to live "when they got tired of California" and returned to New Mexico.

In some Spanish-American villages Carlson found that more than half of the housing units were new mobile homes. It was ironic that Anglo newcomers wanted to renovate and preserve the picturesque old adobe houses, whereas young Spanish-Americans preferred the creature comforts of modern mobile homes.

Jackson said that the mobile home was ubiquitous in the cities, towns, and remote villages of northern New Mexico, and he marshaled a lengthy catalog of common aesthetic criticisms. It is a low-cost, mass-produced, disposable industrial product with no signs of craftsmanship. It has an ugly boxlike shape, it is awkwardly laid out, and it is small and cramped. "Almost from the first day of occupancy," he said, "it spills its contents—and its occupants—into its surroundings: parked cars, refrigerators, packing cases, children and dogs and laundry invade the landscape."

Then Jackson argued that the critics miss their mark. "The villagers who have moved into trailers are in general satisfied," he said. "They may regret leaving the old adobe house, but it is a joy to move into a brand-new house, clean and never used. . . . It takes only a few days to realize how convenient and comfortable a trailer is, and how easy to maintain. . . . For a great many families the trailer is a sensible way of living."

Jackson also reveled in the trailer courts developed for construction and pipeline crews in the Four Corners country, where Utah, Colorado, Arizona, and New Mexico conjoin. They were small and compact, with an average of 25 to 50 units, and the developers wasted no money on beautifying them. "Whenever a new construction project is announced," he wrote,

> the construction company or a promoter (usually operating on a shoe string) leases land from a rancher out where a camp would be profitable and convenient. The promoter drills for water, bulldozes a main street and a few laterals, digs septic tanks, hitches up to the power lines, and builds a public toilet out of cement blocks. But he is not ready for business until he builds the cement block building which is to house the camp laundry. This is the basic social institution; it serves as a gathering place for the women, it is where mail is distributed, notices are posted, and where the pay telephone is installed.
>
> When the boom slackens, when the construction projects are completed, the trailer courts start to shrink and finally vanish. All that remains are patches of brown grass, once lawns three feet wide.

Mercer County, North Dakota

Caroline Tauxe was appreciably less enamored of the trailer courts for construction workers that she found in Mercer County, North Dakota, which had an energy boom between 1973 and 1983. The construction and expansion of four large lignite-coal strip mines, four coal-burning electricity-generating plants, and a coal gasification plant attracted a temporary labor force that peaked at 5,854 workers in 1983.

Many of these workers were specialized construction craftsmen—boilermakers, carpenters, electricians, ironworkers, pipe fitters—who migrated in their mobile homes from one major construction site to another. They rented spaces in trailer courts, transient encampments developed on short notice and lacking any residential amenities. "In the trailer court where I lived in 1978," wrote Tauxe, "the mobile homes were set up on concrete blocks on the raw red dirt, which tracked everywhere in muddy or dusty profusion, depending on the weather. There was an uncommitted, temporary feel to life there, and little sense of community."

One girl of 11 who lived in this trailer court with her family said that she could remember 21 moves, but Tauxe also learned that living in their own mobile homes gave nomadic families a comforting element of stability and continuity. One boy said, "The trailer is nice. It is tough to be always in a new school, to always be an outsider, but year after year I have always slept in the same room at night, no matter where it happened to be."

Furthermore, some of the workers already knew each other from having worked together on previous jobs. Workers skilled in specific crafts are needed during specific phases of large construction projects. They are laid off when that phase has been completed, but they simply move to a new construction site where they may encounter some of the same people with whom they have been working.

Tauxe accused local planners of practicing class-based discrimination against trailer parks and trailer people. Before the construction boom, mobile homes had been in mixed residential neighborhoods, but planners used strict zoning regulations to segregate the temporary work force in new marginal trailer parks. The "better" citizens wanted no part of trailer parks, and one councilman expressed the popular mood when, in response to a request for a zoning variance to develop a park next to a landfill site well outside of town, he opined that right next to the town dump seemed like the perfect place to put a trailer park.

Upstate New York

Janet Fitchen found similar "class-based" discrimination against trailers and trailer people when she investigated social, economic, and political conditions in rural communities in 15 upstate New York counties during the late

1980s. In affluent areas of major second-home development, the new "city people" who lived in "architect houses" deemed themselves superior to the local people who lived in trailers. The affluent newcomers advocated strict land-use regulations to prevent excessive development and to eliminate the "visual pollution" of mobile homes.

Townships with high housing values and affluent residents had enacted highly restrictive ordinances against mobile homes, or even banned them completely, even though mobile homes were the only type of housing that many low- and moderate-income families could afford. Furthermore, few young families could afford to buy land to set up a trailer in popular second-home areas, where prices ranged up to $3,000 an acre and higher, and regulations that required minimum lot sizes of five acres even for a trailer put the price of a home still farther out of reach. "Even if property is available within the family," Fitchen said, "people are no longer free to put their own trailer on it because of tightened land-use and building restrictions."

Exclusionary restrictions on mobile homes in affluent townships had forced young families to seek homes in other areas where pockets of poverty have persisted for decades, areas with "small clusters of dilapidated housing: run-down farmhouses, very old trailers encased in wood additions, shacks built of used lumber and tar paper, converted school buses. Space around the houses is strewn with old cars and car parts, building materials, ancient appliances. . . . When the children of these pockets of poverty grow up, most settle close to home, often in a cheap old trailer placed in the side yard."

Informal roadside trailer clusters were sprouting up in many locations, typically two to four old units sitting in a yard next to an old farmhouse. A group of three or fewer trailers did not come under much regulation or licensing as a trailer park, and three units could share a single septic system.

The edge of nearly every village had a trailer park or an informal cluster of trailers. Every park was full, and the rents for spaces kept rising because the demand far exceeded the supply. Crowding was increasing—crowding of trailers in the parks, and of people in each trailer. Most residents would have preferred a single-sited unit on a lot they owned, but few could afford the costs of buying land, drilling a well, installing a septic system, and moving the trailer.

Many old trailers still occupied by low-income people predated the imposition of even minimal safety standards in construction and materials.

They were potential firetraps, inadequate in quality and in space, unbearably hot in summer, expensive to heat in winter. For some young families, however, even a battered old trailer was a step upward, truly an "upwardly mobile" home, because it offered more space and more privacy than living with parents or in-laws.

Adams County, Wisconsin

Adams County, Wisconsin, is a low-cost summer resort area, and mobile homes are a major component of the housing stock (Fig. 30). It is on the Central Sand Plain, which is notorious as the poorest part of the state south of the boreal forest. Alex Richter, the county extension agent, said that much of the land in Adams County could have been bought for as little as $50 an acre, or even for back taxes, as late as 1960. In fact, the land was so cheap that the Nekoosa Paper Company did buy 36,000 acres, mostly in the northern part of the county, and manages it to produce pulpwood and sawtimber on a 45-year cycle. The company's lands teem with white-tailed deer, ruffed grouse, and other forms of game.

The trailer tradition in Adams County dates back to at least as early as 1926, when Irwin D. Linehan and three other Chicago entrepreneurs decided to capitalize on cheap land by developing a "poor man's recreational paradise" at Dellwood on the banks of the Wisconsin River at the western edge of the county. They laid out streets and riverfront parks, platted 4,700 lots measuring 50 by 125 feet each, built a hotel, a community house, and a dance pavilion, and brought prospective customers from Chicago on special weekend excursion trains with a round-trip fare of $5.25.

The lots at Dellwood originally sold for $49.95. They did not seem small to people from densely built-up urban areas, low-income people who could not afford land in more traditional summer resort areas. Many of these people could not even afford to build conventional summer cottages, so they wheeled in trailers and set them up on blocks on their lots. As late as 1972 a survey found that mobile homes comprised 20 percent of the housing stock in Dellwood, and that percentage does not seem to have decreased subsequently.

The lots at Dellwood were not intended for year-round use, and many were occupied for only two to four weeks in summer plus the odd long

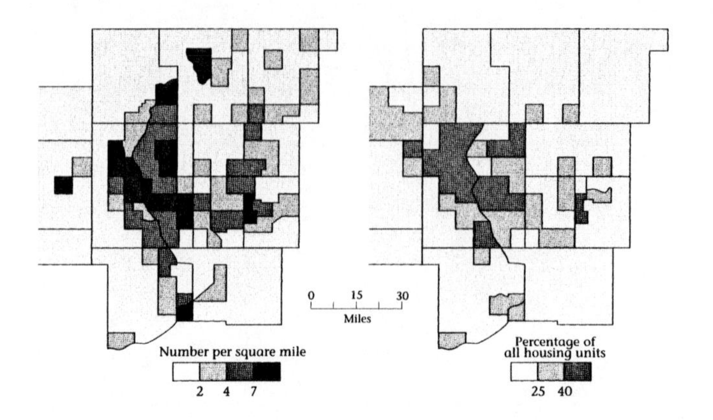

Fig. 30. Mobile homes outside incorporated places in central Wisconsin, 1990

weekend the rest of the year. Despite such brief occupancy, however, the summer cottage, even if it is only a trailer, becomes an integral part of the family tradition. Many of the original owners, and their children and grandchildren, have kept coming back to it, summer after summer, no matter what its flaws. It might be upgraded, as many have been, but "going back to the cottage" is a summer ritual, sometimes even a sacrifice, that exercises an almost mystical hold over the family even when they have become able to afford a far better place.

Most of the lots at Dellwood had been sold by the time of the stock market crash in 1929, which put an end to development in Adams County until after World War II, but the people who kept returning to Dellwood helped to publicize the county as an inexpensive summer resort area for low-income people, and they helped to legitimize the trailer as a summer cottage. Adams County become known as an area where land was cheap and trailers were acceptable.

After World War II better roads and better automobiles made the county more easily accessible from Chicago, from Milwaukee, from Madison. Its attractiveness as a summer resort area was enhanced enormously when some of the paper and utility companies upstream cooperated to build two large hydroelectric power dams on the Wisconsin River. In 1950 they completed Petenwell Dam, which impounded the second largest lake in Wisconsin, and in 1951 they completed Castle Rock Dam, which impounded the state's fourth largest lake.

The Petenwell and Castle Rock dams translated the western border of

Adams County into an almost continuous sheet of water, which boasts two state parks and five county parks with extensive campground areas. Access to the lakes is easy even for people who do not live on their shores, because they have many public boat-launching ramps and sandy beaches. In addition to the usual water sports, the lakes offer waterfowl-hunting areas that are popular in the fall.

The blue-collar image of Adams County might have dissuaded well-to-do visitors, but since World War II the county has attracted a steady summer influx of people with limited financial resources. They could afford to buy land in the county, and modest summer homes are scattered across it on lots of two to five acres.

Many of the new summer homes are small owner-built cabins without running water or electricity. Although the houses are modest, their owners take great pride in their appearance and upkeep, and they have done whatever they could do with little cash but great expenditure of time and effort. The buildings are neatly painted, and they are surrounded by extensive mowed lawns adorned with flowers and shrubs. People who live on cramped lots in the city seem to like large lawns, and riding a power lawnmower is apparently one of the more popular outdoor activities in the county.

For many owners, hauling in a mobile home was an attractive alternative to building their own cabin. It was quicker, easier, and just as inexpensive, and it carried no social stigma in an area where mobile homes comprised nearly half of the housing stock. Many owners have modified their units extensively by building on decks, new rooms, and other additions. A garage large enough for two cars, or for a car and a boat trailer, often dwarfs the dwelling itself.

Adams County has few restrictions on mobile homes. Dennis Hoeffle, the county planning and zoning administrator, said, "Our 3 zoned townships have areas designated for mobile homes, but in the other 14 townships our only control is sanitary. You have to hire a licensed installer to put in a septic system, but that's it."

The county has relatively few mobile home parks, and most units are on individual properties. The town of Big Flats (townships are called "towns" in Wisconsin) in north central Adams County might have the densest concentration of single-sited mobile homes in the United States (Fig. 33). The cluster of houses at the center of the town, where State Highway 13 intersects

Fig. 31. Camper trailer on a private lot in Adams County, Wisconsin

Fig. 32. Single-wide mobile home with added deck in Adams County, Wisconsin

County Road C, was wiped out by a tornado in August 1994, and many owners have replaced their demolished houses with mobile homes instead of rebuilding them.

Some sections in the town of Big Flats have no houses at all because they are in the hands of strong owners, such as the Nekoosa Paper Company, who do not need to sell land, but some rural roads are lined with mobile homes interdigitated indiscriminately with conventional houses. All of these mobile homes are long, narrow single-wides, with no multisectional units, and many are in less than pristine condition.

Roger Fenzaw, who owns Pop's Welding Shop in the western part of the township, said, "This is just a depressed area, and people are willing to sell land dirt cheap. You can buy it for $1,000 an acre and up, and you can buy a used trailer for $500 and up. The inspectors come in and tell you that you've got to plant it or move it. Taxes are low, and this is a good weekend getaway from Madison, Milwaukee, or Chicago. It's a big hunting area, with lots of deer and turkeys, and a lot of the trailers are used for hunting shacks."

Dan Fenster figured that about half of the mobile homes on his road were seasonal, but half were occupied year-round. "Some of us have retired here," he said, "but there are lots of young couples, and they drive to jobs all over the place. I guess you could say that the gal that lives in that one over there is a kind of professional barfly. I think she works as a topless go-go dancer when she needs money. She moved in there after she broke up with her husband. She's had a whole series of live-in boyfriends, and her last boyfriend still lives there. She went out West, Las Vegas or someplace, when she busted up with him, but I guess maybe she'll come back one of these days."

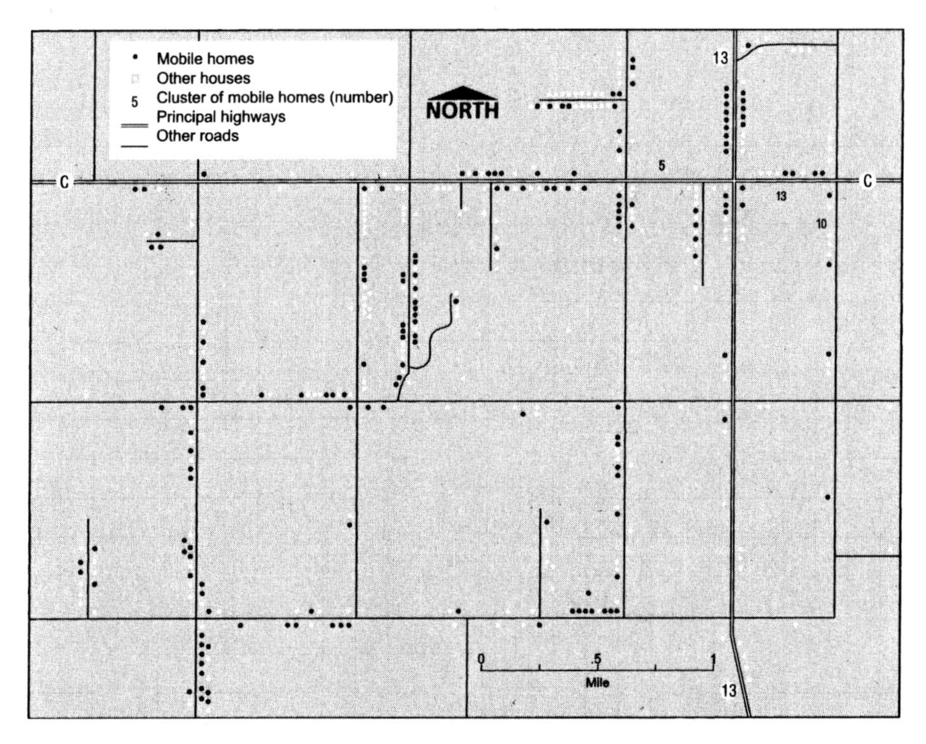

Fig. 33. Types of housing units in part of northern Adams County, Wisconsin, 1996. Big Flats is at the intersection of State Highway 13 and County Road C.

Mobile homes are inexpensive housing in an area that needs it badly, but they also enable their residents to spend their money on more important things, such as cars and boats and guns, and even when they are immobilized they imply a sense of mobility and freedom. Randy Bakovka, who owns the Pineland Camping Park, spoke for many others when he said, "I live in a mobile home. I don't want to have $100,000 tied up in a house that I might have trouble selling if I decide to pack up and move someplace else." It is paradoxical that the people who use their mobile homes seasonally may be more permanent than those who live in them year-round.

Many mobile home owners, both seasonal and year-round, have upgraded their units by building additions to them, and a few have completely replaced their old mobile homes with conventional site-built houses. Adams County, which for so long was synonymous with rural deprivation, has started to show signs of good fortune, albeit hardly affluence. Few areas in the state of Wisconsin have been transformed so impressively in the last two decades.

In the 1970s and again in the 1980s, Adams County had the second highest population growth rate of any county in the state. The boom of the 1970s centered in the town of Rome, the northernmost township in the county, where a developer impounded three new lakes and sold 6,000 lakeshore lots. Most of these lots have conventional houses that are year-round residences for people from the paper mill towns just upstream on the Wisconsin River or summer cottages and retirement homes for people from more distant places.

The boom of the 1980s was more general, and the county enjoyed a steady influx of people of all ages, not just retired people. Alice Parr, executive director of the Adams County Chamber of Commerce, said, "People come here for vacations, and they like it so much that they decide to move or retire here. They want to get away from the crime and crowding in the cities, and they can afford to live here. Lots of young people are moving here, and our school enrollment is way up."

In the early 1980s the county's business patterns showed few signs that it was one of Wisconsin's leading summer resort areas, perhaps because low-income visitors did not spend much money but even brought provisions with them from the city. The prevalence of large ice machines in front of nearly every business establishment identifies the only item that could not last for a visitor's entire weekend stay. By the mid-1990s old structures had been spruced up, shiny new structures had gone up beside them, and the county had more and better motels, eating places, shops, and other businesses that obviously catered to the tourist trade.

Mobile homes remain a major component of the housing stock in Adams County, but one must wonder whether they will become stigmatized as the county sheds its past and becomes ever more mainstream.

Spersopolis

The mobile home is ubiquitous in the rural South, where it is considered a perfectly acceptable form of affordable housing. It has replaced the tumbledown shack of the plains and the log cabin of the wooded hill areas, and often it stands where once they stood to take advantage of the wells from which they drew their water. Inhabited log cabins, which were common in the hills of Appalachia only a generation ago, are hard to find today because they have been replaced by mobile homes, and the only remaining inhabited log cabins are those that have been gentrified by yuppies.

As in other parts of the country, many a farmhouse in the rural South has a single-wide mobile home in its side yard. It is a starter home for the married son who joins his father in the farm operation. Eventually, when the parents grow older, the two families swap houses; the son and his family move into the farmhouse when he takes over the farm, and his aging parents retire to the mobile home, which is smaller and easier to maintain.

The rural mobile home attains its apogee in the new manufacturing belt of the Southeast, which sprawls across the Piedmont of the Carolinas and Georgia and the Great Valley of northeastern Tennessee and southwestern Virginia. Ten percent of the nation's mobile homes in 1990 were within a 250-mile radius of Charlotte, North Carolina, and most of them were single-sited on individually owned highwayside lots.

A new pattern of dispersed settlement has evolved spontaneously in this new manufacturing region, which we have dubbed "Spersopolis" for want of any generally accepted colloquial name. Spersopolis is our closest approximation to the ideal populist city of the automobile age, the way most Americans would elect to live if they had complete freedom of choice, although many other parts of the nation show similar trends.

Fig. 34. Single-wide mobile home on the site where a log cabin once stood

Fig. 35. Mobile home park near Winterville, North Carolina

In Spersopolis the central cities retain their traditional nine-to-five office functions in shining new towers, but visitors often complain that their downtowns are deserted because most of their major retail activities have fled to suburban malls, and downtown is little more than a convenience shopping area for those who work there. Many of the new factories are on inexpensive land in distant, low-tax rural areas.

The residential areas of Spersopolis sprawl far into the countryside, because long-distance commuting has become the preferred new lifestyle in an area where roads are good and winter driving holds few terrors. Planners fulminate about the expense of providing water, roads, and other services to a dispersed rural population, but ordinary people like to live in the countryside on their own land.

The growth of manufacturing in Spersopolis has coincided with a decline in agriculture, which has released an abundance of low-wage workers for the new rural factories, plus an abundance of low-cost land that enables these workers to live as and where they wish. Many prefer to live on their own land in rural areas instead of moving into town, perhaps in reaction against the grim regimentation of the early textile mill villages, although attachment to place and ties of kinship are also important.

The geographer Alfred Stuart says that the textile mill companies have offered to sell nicely rehabilitated company houses in their villages to workers at very favorable prices, but the workers prefer to live in mobile homes and spend their money on cars and boats. The workers counter that their mobile homes are brand new, everything in them is new, and everything works.

Farm families have moved from their old homes on unpaved back roads to modern new homes on the blacktopped highways for ease of commuting, and the availability of inexpensive land has lured city folk to the countryside. An almost continuous necklace of houses of every kind lines virtually every paved road except limited-access highways (Fig. 36).

Few houses are more than a few hundred yards from a paved highway, and few stretches of paved highway are more than a few hundred yards from a house. Their highwayside location is well suited to commuting to factories where layoffs are common and employment is insecure. The worker who is laid off at one plant can easily commute to another, and one location is about as good as any other if spouses have jobs in different plants.

Mobile homes are a major component of the housing stock of Spersopolis. They are scattered along the highways in what might best be described as a roadside random distribution, interspersed with other houses of every possible sort. In unembarrassed proximity they sit cheek by jowl with modern suburban-style houses, tumbledown shacks, and mansions whose spacious lawns are maintained better than golf course fairways.

Mobile homes are the residences of choice, or at least the starter homes, for many people whose education, skills, and income are limited. Many families have developed small clusters of mobile homes in "hamlets" around the initial home. The parents let their newly wed children put their mobile homes on the family property, and the grandparents provide conve-

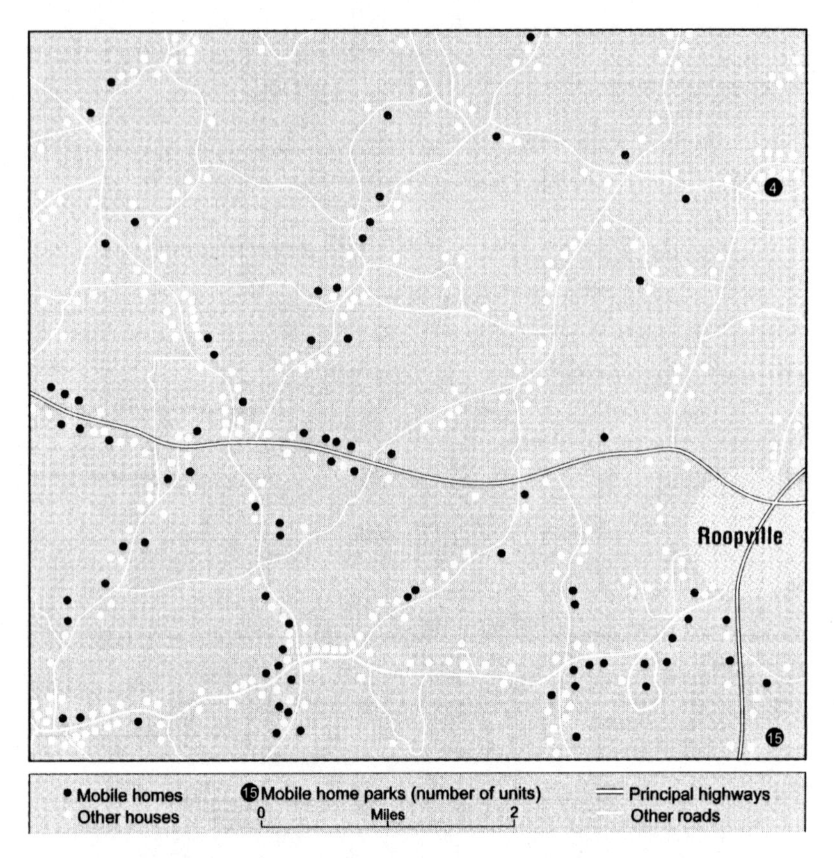

Fig. 36. Types of housing units in part of southern Carroll County, Georgia, 1996

nient day-care and baby-sitting services for the young couples, who both must work.

The Coal Field

Mobile homes are also a major component of the housing stock of the southern Appalachian coal field of southern West Virginia, eastern Kentucky, and southwestern Virginia, one of the poorest parts of the nation. A new mobile home can be such a symbol of prosperity that the double-wide mobile home has become the model for a new type of structure that Charles Martin called a "half-house." A half-house is roughly 12 feet wide and 30 feet long, with two rooms under a shed roof and windows only on the front wall.

Martin said a half-house is a homemade version of half of a double-wide mobile home. It sits on land donated by the family and can be knocked together over a weekend with family labor by a young couple who cannot even afford to buy a mobile home.

We talked to the residents of 26 mobile homes in the coal field area of Buchanan County, Virginia, to find out who they were and why they lived in mobile homes (Fig. 38). This county is one of the poorest in Virginia. In 1990

Fig. 37. Mobile home park across the tracks from a new manufacturing plant in Atkins, Virginia

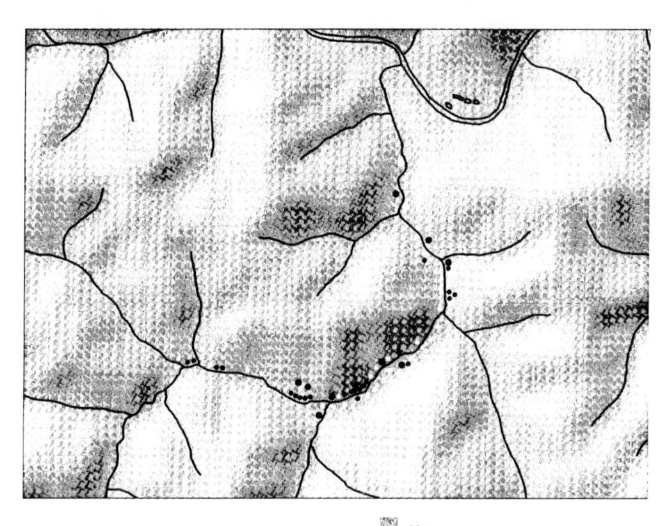

Fig. 38. Types of housing units in the Conaway valley, southern Buchanan County, Virginia, 1998

it had the highest percentage (35.5 percent) of mobile homes in the state, and more than 95 percent of the building permits issued in the county during the 1990s were for mobile homes. We have identified the mobile home residents by letters, according to the order in which we talked to them.

Ms. A (mid-30s) lives alone in a single-wide on land that her mother owns. This is her first home. She needed to be near her sick mother, and she wanted to be close to work. There were no houses in the area that she liked, and she was pleased to be able to pay cash for her new home.

Mr. and Mrs. B (mid-30s) and one child live in a double-wide on land they rent. This is their first home. They had rented an apartment but moved here to be near her family. They could not afford to buy a house but were able to take over payments on the mobile home from a family that was leaving.

Mr. and Mrs. C (late 20s) and two children live in a single-wide on land they were given by her family. They were married young, and this is the first home they have owned. Only the husband is employed, and the wife is attending college. They moved here to be near their families. They did not have the money for a down payment on a house but did not need one to buy a mobile home.

Mr. and Mrs. D (early 20s) and one child live in a single-wide on land they rent from her family. They were able to buy the mobile home without having to make a down payment and are saving money for a down payment on a house.

Mrs. E (mid-40s) lives in a single-wide on rented land squeezed in between two houses. She is recently divorced and was able to buy this mobile home with her divorce settlement. She wanted to move away from her ex-husband but still be close to her place of employment.

Mr. and Mrs. F (early 20s) and the new baby live in a single-wide on land next to her parents' house where they do not have to pay rent. They financed their mobile home through a dealership because they did not have the money for a down payment, had not established credit, and so could not qualify for a home loan.

Mr. and Mrs. G (early 30s) and two children live in a double-wide with a large attached deck on land they own. Previously they lived in a single-wide on the same land. When they started looking for a new home they found they could buy a very nice double-wide for a lot less than the cost of a new house. It is well built and sturdy and has plenty of room for the entire family. They used the single-wide as a trade-in and had very small payments.

Mr. H (mid-20s) lives alone in a single-wide on land he rents from his parents. Being near them is important to him, and he is also near his job. He wanted to buy a house but could not realistically afford one.

Mr. and Mrs. I (mid-20s) live in a single-wide on land they rent, but they expect to buy land soon. They both work full time, and both attend college in the hope of getting better jobs. Previously they lived in an apartment, but the mobile home dealer gave them a financing option that cost them less per month than they had been paying in rent.

Mr. and Mrs. J (late 20s) and one child live in a single-wide on rented land, but they hope to buy land soon. They wanted to be near his job but could not find a house in their price range and bought this secondhand mobile home from relatives who were moving.

Mr. and Mrs. K (early 70s) live in a double-wide on land they own. They sold their house, which was too big for just the two of them, and paid cash for their new mobile home. Neither is able to drive, so they need to live near family who can give them rides to the grocery store, the doctor, and so forth.

Mr. and Mrs. L (mid-30s) and three children live in a single-wide on land they rent, but they hope to move it to land they have inherited. They both are disabled and have limited income, but the dealer arranged a payment schedule they can afford.

Mr. and Mrs. M (early 20s) are newlyweds who live in a single-wide on land they rent. They do not intend to stay in this area and did not want to invest a lot of money in a house. He is already looking for a job elsewhere, and their mobile home is for sale. They figured it would be easy to sell when they moved.

Mr. and Mrs. N (early 40s) and their two children live in an especially nice large triple-wide on land they own. Previously they owned a double-wide that they sold to buy the triple-wide. They had looked at houses but could not find one they liked within their price range, and the triple-wide was twice as nice as any house they could have bought for the same price.

Mr. and Mrs. O (mid-30s) and one child live in a double-wide on land they lease really cheaply from the railroad. Previously they owned a single-wide. They had wanted to buy a house but could not afford the large down payment the bank wanted before lending them the money to buy it, and had been able to special-order their mobile home and pick out everything they wanted in it.

Mr. and Mrs. P (mid-50s) live in a single-wide on land they rent. They live here part of the year and in Florida part of the year. They sold their house and moved into the mobile home because the house required too much upkeep.

Mr. and Mrs. Q (mid-40s) live in a single-wide on land they own. He was laid off for an extended time a couple of years ago, and they had to sell their house and move in with her parents. When he was employed again they saved their money for a down payment to buy this mobile home and the land, but they are thinking about moving someplace where he can find a better job.

Mr. and Mrs. R (mid-40s) and one child live in a double-wide on land they own. Previously they owned another double-wide. They did not want to take on a large amount of debt, because he is laid off frequently, and they put most of their money into a pet store she owns in town. Several family members live in adjacent mobile homes, and she said they like living close by.

Mr. S (late 30s) lives alone in a single-wide on land he rents. His wife got the house after their divorce, and he cannot afford a house because of support payments. He would not have been able to afford even the mobile home if the dealer had not worked out a low schedule of payments for him.

Mr. and Mrs. T (mid-40s) and two children live in a single-wide on a tiny lot they rent on the side of a steep hill. Previously they rented a mobile home, but this is the first one they have owned. They want to stop renting land because they feel like they are throwing money away, but they are having trouble finding land they can afford to buy.

Mr. and Mrs. U (mid-20s) live in a single-wide on land his parents gave them when they married recently. They do not need a large home at this time, and they wanted to live near his parents. They also like the modest down payment and the low monthly payments on their home.

Mr. and Mrs. V (mid-30s) and their three foster children live in a double-wide on land they own. Previously they owned a single-wide. He is in a manager-training program at the store where he works and might be relocated when he completes it. They saved most of the price of their home and financed the rest through a bank because the dealer charged interest rates that were too high.

Mr. W and his girlfriend (early 30s) live in a single-wide on land he owns. He said he did not want to have to mess with banks when he traded in their old home for the new one. The dealer fixed up everything for him and saved him a lot of headaches.

Mr. and Mrs. X (early 20s) and two children live in a single-wide on land they rent. They were able to buy the home because a down payment was not required. They were married right out of high school, and neither has established credit, although both work full time. This is their first home, and it is important to them to own it instead of renting it.

Mr. and Mrs. Y (mid-20s) live in a single-wide on rented land. They bought their home from a couple who were moving out of the area. He is working full time, she is working part time, and both are attending a community college. When they complete their studies they plan to move to a four-year college to complete their education, and they will rent or sell the home when they leave.

Mrs. Z (early 60s) lives in a double-wide on land she owns. She could not maintain their large house after her husband died, so she sold it and bought this home and land to be near her children.

PART THREE

SIDE BY SIDE

The Mountain West

The Mountain West relies more heavily on mobile homes for its housing stock than any other major region of the United States (Fig. 25). We selected Flathead County, Montana, for a closer look at mobile homes in this region. Advertising agencies set out for places like Flathead County to shoot commercials for sport-utility vehicles and light beer. Glacier National Park and the Swan Mountains, on the county's eastern edge, invite adventure-seeking travelers into the heart of wilderness, much of it accessed by logging roads and well-groomed wood-chip and gravel trails.

Flathead County, some would say, has been "discovered," inundated by tourists and migrants alike who are attracted to Big Sky, small towns, and low land costs. Over the past three decades both the county and its towns have grappled with a flood of new residents. The population of Kalispell, the county seat, grew 19 percent during the 1990s, and that of Whitefish, 15 miles to the north, grew 15 percent. Planners expect the county's population to keep growing, swelling to upwards of 97,000 by the year 2010. Currently, almost 75,000 people call it home.

Over the past three decades, northwestern and west-central Montana have gone from being the middle of nowhere to being the end of the line for an increasingly mobile western population. Many of the new residents up and left their homes elsewhere in Montana, hoping to find work or attend school in Kalispell or Missoula. In towns such as Bozeman and Helena, where the arid shortgrass plains give way to a more moderate mountain climate, locals brace themselves for continued migration from both inside and outside the state.

More often than not, the new Montanans (and Idahoans and Eastern Washingtonians) are renting their U-Hauls on the West, rather than the East, Coast. Folks are leaving Los Angeles, San Francisco, and Seattle, escaping the crowded freeways and high costs-of-living that result in part from previous generations of people seeking "the good life." In 1994, after yet another major earthquake hit the San Francisco area, realtors in western Montana were inundated with phone calls inquiring about available land. The disaster was for some the last straw—leave the city now and head for greener pastures.

During the 1980s and 1990s, rural counties in the Intermountain West, between the Cascades and the Rocky Mountain Front, grew faster than rural and urban counties in the rest of the United States. Much, if not most, of this growth can be attributed to the movement of urbanites into the small-town and rural fringe away from the big city. Some towns—Sun Valley, Moab, Colorado Springs—transformed themselves almost overnight. (Some would argue that others changed them.) Most had a slower transition. The resulting culture clash between younger, educated amenity-seekers and longtime blue-collar resource-industry workers means that Main Street is now home to both cowboy paraphernalia and cappuccino stands. Missoula, Montana, may well have more coffeehouses per person than the Starbucks meccas of Seattle or Vancouver.

The region's growth provides North Americans with a mirror of their lifestyles. Their shopping habits are reflected in the construction of new Wal-Marts and strip malls, the popularity of gourmet and ethnic restaurants and food stores, and a growing number of fast-food joints. In the summer, tourists may stop in any of these places, but the locals keep them running year-round. Not all of the new businesses are chain operations. Downtown cores are filled with entrepreneurial ventures still in their infancy, started by regional artists and transplanted West Coast executives and middle managers.

A booming housing industry has reflected and celebrated the arrival of new residents, and is helping to alter the western landscape, sometimes to the chagrin of local residents. In high-amenity, less populated areas outside of resort towns and small cities in the West, the demand for ranchettes—parcels of 20 acres or less—has resulted in the rapid subdivision of the rural landscape. "Monster" homes and exclusive condominium complexes carve up lakeshores and riverfronts. Inevitably, these developments raise the ire of residents, in part because of the loss of access and "ruralness," in part because the people moving in are outsiders, in part because locals themselves could not afford such grand homes.

More significant yet less offending are the small starter homes, new subdivisions in town, and manufactured housing. While Hollywood's best known may be building palatial cedar and stone remote mountain hideaways, the majority of new and old residents alike stick with the more modest options available to them: older homes in town, newer homes outside, and mobile homes everywhere.

Filling in the Flathead

It is a familiar early-morning sound. At 7:20 A.M. the Empire Builder cautiously creeps up on the Great Northern Depot in Whitefish, coming to rest alongside the loading platform. After a short break, Amtrak's passenger service continues east, eventually reaching Chicago two and a half days after leaving Seattle. In 12 hours westbound passengers will gather at the depot to board a different Amtrak train leaving for Libby, Sandpoint, Spokane, and points beyond.

Whitefish, Columbia Falls, and Kalispell—the three towns that make up Flathead County's tri-city hub—grew up around the railroad. Whitefish still depends on it for tourism. In Columbia Falls the aluminum plant relies on the freight lines for receiving and shipping goods. Kalispell also uses rail for shipping, and much of the downtown core straddles a single set of tracks.

As in much of the West, placement of the railroads was crucial to the success or failure of a particular town. Flathead Valley's first settlement, Demersville, quickly folded after it was decided that the Great Northern would pass a few miles north of it. Townspeople relocated, building the foundations of what would become Kalispell. Trains brought more folks in to settle the expansive and lush valley floor, and cattle, agricultural goods, and timber products were sent out. Within two years of the building of the railroad, the region had a sufficient number of new residents to warrant the creation of a new county. In 1889 politicians carved Flathead County out of Missoula County.

The key to the county's growth has always been its reliance on more than one resource. Some places began as single-industry towns (Whitefish was nicknamed Stumptown in celebration of its logging heritage), but the county's residents as a whole worked as loggers, ranchers, farmers, merchants, and government officials. As early as 1910, with the creation of Glacier National Park, the area also benefited from tourism. Flathead County weathered the state's economic turmoils—from the Depression to Butte's fall from grace—better than most places, in large part because it had a far more stable economic base.

Kalispell blossomed as the heart of northwestern Montana. Although other towns in the county occasionally had little or even negative growth, Kalispell continued to expand, and the town has not lost population in any decade since 1920. Whitefish, on the other hand, lost, then gained back, then lost again, hundreds of residents up through the 1950s. Since that time it too has continued to lure new residents, most of whom were more interested in playing in the forests than in working in them.

Kalispell also benefits from a wealth of two types of amenities—historic and natural—that appeal to residents and visitors alike. Kalispell strives to remind the visitor that the town holds on to its western culture. False-front saloons, multistory brick buildings with apartments over stores, and cowboy-boot-and-hat-motif window displays line the commercial stretch of Main Street. Two-story homes, many of them on the National Register of Historic Places, make up most of the housing in a six-block radius of Main. The east side of the city was developed first and is the site of the Conrad Mansion, home to the wealthy businessman Charles Conrad at the turn of the last century. Most houses near the downtown are from half to three-quarters of a century old.

Contrast this vintage scene with Idaho Street, the section of Highway 2 that is the other major traffic corridor in the city. A smattering of hotels, fast-food restaurants, grocery stores, gas stations, and small strip malls lure the Glacier-bound traveler in for a quick meal and supplies. The "big box" buildings, housing Tidyman's and Fred Meyers, provide goods for the local population. Kalispell has two malls, one of which is in the heart of downtown. Further east on Highway 2 beyond the city limits, in the unincorporated area known as Evergreen, a number of discount and warehouse stores now compete with Kalispell's downtown businesses for customers, and they do not have to pay municipal taxes.

The small city also boasts a variety of natural and scenic assets. Kalispell developed near the junction of three major northwestern Montana rivers:

the Whitefish, Stillwater, and Flathead. By late spring the rivers are swollen with snowmelt from all but the highest elevations of the Whitefish and Swan Mountains, which occasionally causes problems for areas south of Kalispell, but flooding is far more common and severe in the Evergreen area. Traditionally land has been available at a lower price on the flood plain, but prices have risen as developable land close to and within Kalispell becomes harder to find. For a few weeks every year, then, Evergreen residents living on the flood plain cross their fingers, bite their nails, and hope for a cool, dry spring.

Because of Kalispell's location in relation to the rivers, the Swan Mountains, and Flathead Lake, most new residential developments come complete with at least some sort of view and a wealth of recreational opportunities. On the rolling hills north of town and the flat valley floor to the south, residents also benefit from living next door to small ranches and farms, the ideal rural-idyllic: horses put out to pasture, rows of rustling willows and alders along the creek beds and rivers, and aging barns of warped, weathered boards held together by rusted nails.

These views are guaranteed, in part, by the federal government, Flathead County's largest landowner. Together, all levels of government own more than three-quarters of the land in the county. Much of this land is protected, either as part of Glacier National Park, as recreation and wilderness areas in the Swan and Whitefish Mountains, or under wild-and-scenic-river status. Most of the rest is Forest Service land.

Government ownership of the land, little of which includes the valley floor, affects settlement in the county in a number of ways. First, it provides a fair amount of employment in Glacier and the state parks, in the Forest Service, and in day-to-day municipal affairs. Second, Kalispell's proximity to protected areas allows for recreation and, to a degree, sets limits to growth in mountain areas. Last, this ownership pattern allows the municipal, county, state, and federal government to consolidate inventories and technologies for planning purposes, most notably Geographic Information Systems (GIS). Flathead County has successfully combined local consultation and federal information in creating long-term planning strategies and goals for its master plan.

Despite the best planning efforts in Flathead County and elsewhere, however, the rapid population growth in the West has outstripped the ability of the average small town or city to accommodate all its new residents. Even

Boulder, Colorado, with its urban growth boundaries and set population limits, must balance the needs of longtime residents and would-be transplants, including providing adequate housing and employment. In Kalispell, Missoula, and Helena, the influx of new residents has meant that some resources are in short supply. Among those that are in greatest demand: urban land and affordable housing.

Population pressures are creating a changing urban setting and an increasingly divided rural landscape. Despite the best efforts of Flathead County planners to concentrate settlement near the urban cores, particularly Kalispell, large portions of land outside zoned tracts on the valley floor are fair game for development. So while much of the area inside Kalispell's zoned area is deemed agricultural or suburban agricultural, land lying just outside the invisible zoning boundary is rapidly being subdivided.

"I came here in 1968, and I have watched it go almost out of control," remarked Jimmie Moore, who along with his family owned a home just outside Kalispell. At the time we talked to him in the spring of 1997, his double-wide mobile and small property were for sale. Moore said that he planned on moving further away from the busy-ness of Kalispell. "There used to be no traffic jams, no nothing. You could go five minutes out of town and hunt a deer. It used to be one of the best little towns in the country."

The West and western Montana are no strangers to dealing with population growth. But the problem of learning how to cope with population *density* is new. While the region's largest cities are old pros at urban planning, the rural fringes and small towns are not. During the 1970s, when people began moving in, housing was in short supply. Rural areas, which traditionally did not provide enough demand for a large local housing industry, relied on mobile homes to fill the gap. Neighborhoods outside the small city grew; property owners began subdividing, and agricultural lands nearest to town were under increasing development pressures.

Mobile homes provided the main solution to housing problems in the Kalispell area and most of northwestern Montana. From the late 1960s on, mobiles filled new subdivisions and parks, taking their place alongside older site-built homes outside city limits. In some cases these mobiles were not just the quickest and most cost-effective option, they were the *only* option. Entire neighborhoods—indeed, entire towns—teemed with single-wides with

Fig. 39. Mobile home in Evergreen, Montana, with an added porch and a chain link fence around the yard

white siding and aluminum shutters. On city streets and old logging roads alike, the changing nature of the region's housing was becoming clear.

Despite the predominance of manufactured housing, neighborhoods were anything but homogeneous. Different styles of single-wides, coupled with the rising popularity of the double-wide and the presence of site-builts in the area, meant that housing along the average street was as diverse in appearance as it was in construction. What's more, owners built on extra rooms and decks, changed external features such as siding and skirting to suit their tastes, and over the years created a whole new home of which the original mobile was only a part. Only mobile home parks, with their uniform lot sizes and home placement, even remotely resembled the cookie-cutter developments of the East and Midwest.

A Snapshot of Evergreen

During the past three decades, no place in Flathead County grew as quickly and dramatically as Evergreen. An area of primarily lush green farmland with only sparse residential settlement before the 1960s, this unincorporated township became a model of an almost cancerous pattern of subdivision. By the late 1960s, rapid development had spread beyond the Kalispell city limits. New residents moved to outlying areas, where land prices and taxes were lower and the views were better. Piece by piece, large agricultural properties were subdivided and sold. Evergreen offered the lowest prices of all, given its site on a flood plain and its preexisting street network. In 1970, 1,000 people called this little place home. By 1980, 3,710 people lived here. During the 1990s the rate of growth exceeded 50 percent, as the population leaped from 4,200 in 1990 to 6,215 in 2000. Evergreen was the fastest-growing place in the state.

In 1965 Harold Johansen and his wife Sarah moved to southeastern Evergreen at a time when the area was still all farmland. "I was the first person in here," remarked Harold. "Not quite the oldest, though." The couple bought a mobile home to live in; nothing else was available. Indeed, their entire street quickly filled with other mobiles. Even today their street has no site-built homes, but other streets have a complex mixture of housing types.

The Johansens were not alone in 1965, nor were they atypical in 2000. Half of the residents of Evergreen live in mobile homes, either in parks or more commonly on their own property (Fig. 40). Although site-built home construction is booming in the Kalispell area, many of the new homes cost more than local folks are willing to spend. In Evergreen, a predominantly blue-collar young-family and retiree area, these new homes also bring with them higher taxes and property costs.

Elizabeth and Larry Parker live in a 1962 mobile home in the most flood-prone area of Evergreen. Since buying the home in the mid-1980s, the couple have made a number of changes to the old trailer, including knocking out a wall and adding a living room and bedroom space. They also hoisted the home onto pillar foundations in hopes of avoiding serious flood damage. Despite the costs of making the changes, the Parkers wouldn't think of living in anything but a mobile home.

"It's the only affordable option if you don't have the money," noted Elizabeth.

"Well, even if you had the money," added Larry, "why pay that much?"

The Parkers' son, John, who lives next door with his children in a single-wide, remarked, "Why pay over $1,000 a year [in property taxes]? That's almost $100 a month. I pay a few hundred a year." In Montana, property

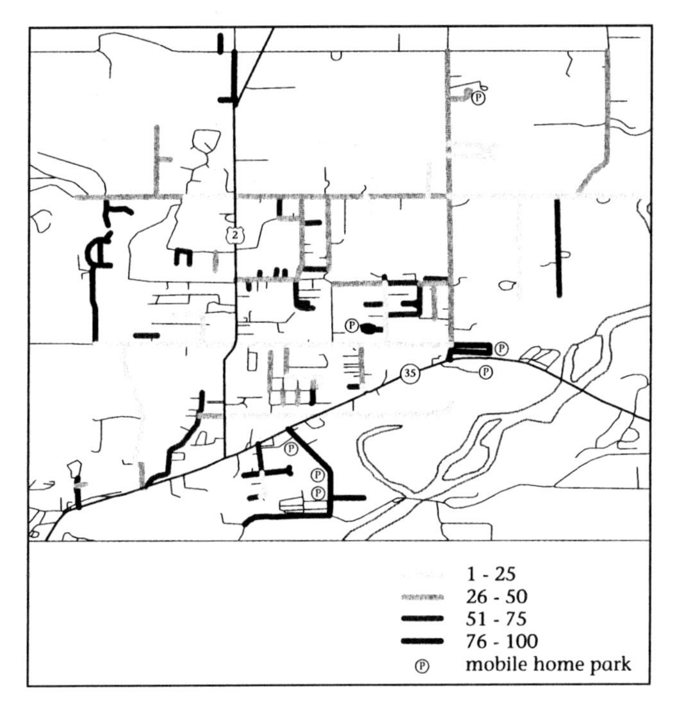

Fig. 40. Mobile homes as a percentage of the housing stock on streets in Evergreen, Montana, 1997

Fig. 41. Double-wide mobile home in Evergreen, Montana, on a solid masonry ground floor

Fig. 42. Types of housing units along a street in Evergreen, Montana, 1997

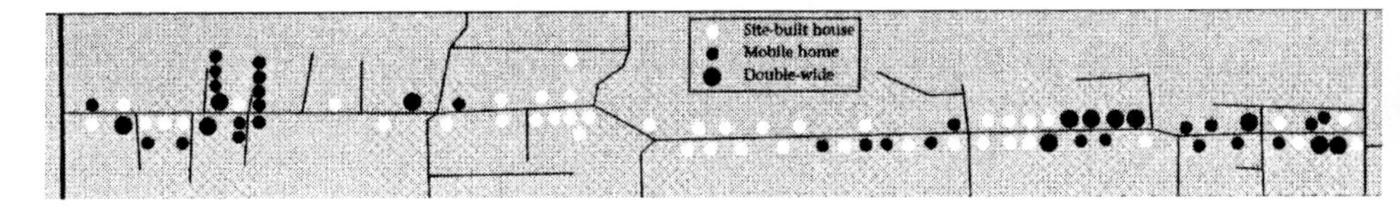

taxes for mobiles remain significantly lower than those for site-builts, even when the mobiles are attached to foundations. Homes are listed according to model and year, given a rating, and taxed accordingly.

The kinds of changes the Parkers made to their home reflect the changing needs of a family as well as the realities of living in northwestern Montana. In the hardest-hit areas of southeastern Evergreen, which is bordered by the Flathead River, almost half the owners loft their homes onto raised foundations in order to avoid flood damage. Any new mobiles moved onto the flood plain must now be put on foundations of some kind, rather than blocks, according to zoning regulations. All over Evergreen, owners of mobiles, particularly the newest models, put their homes on foundations in order to secure loans, guarantee financial appreciation, or make their place more attractive.

In addition to high waters, Flathead County residents also cope with heavy snowfall, which, thanks to a Pacific Northwest influence, is wetter than elsewhere in the state. Manufacturers outfit newer mobiles with higher roofs of 3/12 or 4/12 pitch, but a large number of older homes, primarily single-wides, still have low- or no-pitch roofs. The solution? Build a new roof. These raised roofs may be attached to the home or just built over it. Without the raised structures, many older homes would have had a hard time withstanding the weight of a record snowfall, like the one that struck the area in the winter of 1996–97.

The most common additions, however, are space, not weather, related. Older single-wides in particular provided little breathing room. As families grew, so did the home. Owners added rooms, built decks, knocked down walls, and erected porches and mud rooms. In order to achieve a seamless appearance, these changes often included new siding or masonry, or the construction of enclosed porches, garages, and breezeways.

In northeastern Evergreen, red stenciled angels and cupids adorn Martha and Mark Simpson's brand-new aluminum mailbox. Martha, who does the stenciling both as a hobby and as a job on the side, painted them on the box herself. Twenty feet down the short gravel driveway from the mailbox sits the 1995 white mobile home with blue trim and shutters the Simpsons recently moved into. The yard was still marked by pockets of mud and clay, signs that the home (and the fill it was sitting on) was still the new kid on the block. The couple purchased a pair of single-wides and five acres for themselves and for

Fig. 43. The owner doubled the size of this single-wide mobile home in Evergreen, Montana, by adding a site-built house and garage

Mark's mother, who would live next door, then began a series of year-long renovations. "I still have the stigma on mobile homes," remarked Martha. "I'm doing the inside to resemble a [site-built] home." In the long term they plan to add a deck and build a gazebo in the yard. Inside, Martha put her stenciling skills to good use, adding personal touches to her kitchen and living room.

Down the street from the Simpsons, Rose and Peter Fielding's 1985 Marlette double-wide sits at the front of a long lot reaching back to the river. The home looks surprisingly new, testimony to the care Rose and Peter have given it. A covered patio provides a summertime breakfast area and a bit of storage space for larger items that take up too much room inside the house. The landscaped yard includes a stone fountain built by Peter and a greenhouse. The home has an attached carport, and a chain link fence borders the front yard.

The Fieldings' home and other double-wides like it illustrate how "normal" mobile home living can be, and is. Few owners opt not to make changes or additions to their homes. Given the expense of moving, most owners choose to build on an extra bedroom with the birth of a second child, or a deck for holding parties. Some homes have so many additions that the single-wide mobile is only part of the final product. Referred to as "hybrids" because of their mixed origins, these homes usually have as much square footage in the site-built portion as in the mobile portion, if not more. Hybrid homes look like every other suburban ranch dwelling, usually complete with garage and matching siding or shingles. The single-wide mobile forms one side of the house.

When homes become too small or too old for their needs, Evergreen mobile home owners more often than not choose to replace their home with another mobile rather than move into or build a site-built house. As veterans of mobile home living, these homeowners may move from a single- to a double-wide or from a double to a triple. The newest models of every dimension offer more room and interior storage space than their predecessors, so often a single-wide is replaced with another just like it. Most mobile homes built after 1980, more stylish and designed to last longer than older models, are essentially permanent features of any neighborhood. Many of the homes being replaced are too old to resell to anyone not living in a park, because zoning laws do not allow pre-1972 models to be moved to privately owned lots. (Older homes already in place are grandfathered.)

The newest manufactured homes include so many features of site-built houses that the two types have become nearly indistinguishable, and mobile home living now appeals to more than just the lower-income homebuyer. In places where mobile homes are accepted and common, and where smaller site-built housing is scarce, the newest double- and triple-wides have become a popular choice for middle-income residents. In the Kalispell area, little of the new site-built housing is ranch-style or simple two-story homes. What site-builts are available are in newer cul-de-sac neighborhoods farther from town. These neighborhoods lack the rural feel of much of Evergreen; the subdivision lots are too small to keep horses, too far from small farms, and not near the three big rivers.

The flood plain, population growth, a predominantly low- and middle-income resident profile—all ensure that Evergreen remains ripe for mobile home settlement. Despite the rapid growth of the past three decades, Evergreen has still not run out of room, although the belt is tightening. Much of the new housing, especially mobiles, is on parcels that were divided in half. The original home, which may be either a site-built or a mobile, is almost always at the front of the property near the street. In subdividing, the owner of the land places a second home on the back half. This second home provides a valuable and easily obtained source of rental income, or a quick sale. Many Evergreen neighborhoods are "filling in," so to speak, through this minor subdivision process, which is relatively unobtrusive and is ignored by everyone but the immediate neighbors.

Residents do not greet proposals for large-scale developments with as much nonchalance as they do the minor subdivision of a single property. In 1995 a series of hearings and planning board meetings began taking public comment on a proposed zoning change of a parcel of formerly agricultural land in central Evergreen. The change would have allowed greater densities by shrinking the lot sizes from one acre, for which the area was zoned, to 10,000 square feet per lot. It was not an application for a mobile home park, but the applicants were two of Kalispell and Evergreen's largest manufactured-home dealership owners. The popular assumption was that the newly subdivided property meant additional mobile homes.

The opposition to the zoning change for the most part did not result from a distaste for manufactured homes. Although one resident noted, "The people who move into these manufactured homes don't pay taxes," most concern instead arose from safety issues and the inability of the current road and school network to withstand further population pressure. Others feared the impact of new development on the flood plain and sewer system.

Despite its proximity to Kalispell, many of the residents perceived Evergreen as a rural area. "We moved to a rural area to get out of the city to escape that sort of thing. Now I am being threatened with 350 families moving in across the street on 50 acres," remarked one resident. Another said, "The high density will change the rural density to urban, and that would be a major life change for me." A resident from south Evergreen concurred: "We have four acres and moved there for a rural lifestyle, and feel this is opening the door to too much population density." In the end, despite support for the development from the Kalispell planning department, which saw this project as a way of concentrating growth and controlling sprawl, the zoning request was denied.

The fight over the zone change highlighted a number of problems and tensions surrounding the rapid growth of Evergreen. Residents expressed concern not over the type of homes, but the number. The same factors that lured them to Evergreen—a rural quality of life, lower land values—are now threatened by a flood of people seeking the same thing. The result is increasing development outside of zoned areas north of Evergreen. This development includes monster homes for the rich and cul-de-sac projects and manufactured housing for everyone else. The widening urban-rural fringe places pressure on road networks, rural schools, agricultural lands, and waste disposal systems that are often ill equipped to deal with the new growth.

The threat of annexation, meanwhile, also looms in Evergreen's future. Although Kalispell has succeeded in annexing other outlying areas, many of which welcomed the increase in services, Evergreen has overwhelmingly rejected annexation. This unincorporated area has built its own school and fire systems and does not want to pay taxes into Kalispell's pot. Should Kalispell ever succeed in someday annexing Evergreen, higher lot prices and taxes would follow in short order. Lower-income homeowners, primarily mobile home dwellers, would face tougher financial times, and pressure on agricultural lands beyond the zone boundary would increase.

Boom and Bust in Western Montana

Population did not increase in all northwestern Montana towns as it did in Kalispell and Evergreen during the past three decades. In some places growth was tied to the success or failure of a mill, plant, or project. Regardless of how fast they grew, however, mobile homes have remained the fastest-growing type of housing in almost every town and rural area in this corner of the state.

Thirty years ago the city of Columbia Falls had two and a half times as many people as Evergreen. Today this small community, population 3,645, 15 miles northeast of Kalispell, struggles to stay viable and to attract new businesses and residents. The town relies on the local aluminum plant for jobs and taxes, and the railroad and Plum Creek Timber employ a number of residents. Columbia Falls is a blue-collar town to its very core, one that opted not to jump onto the tourism bandwagon.

As in Kalispell and Evergreen, housing in Columbia Falls reflects periods of prosperity and rapid growth. Older two-story Victorian-style homes from the early 1900s cluster near the aging downtown core. On the other side of Highway 2, trailer parks constructed during the 1960s and 1970s housed a significant number of people flooding the town looking for work, and they remain filled today. Any new homes in and around Columbia Falls are mobiles. Some find a place inside one of these parks; others are plunked onto their own land. In the absence of new, more affluent residents seeking to get away from the big city, there has been little call for large residential developments or the construction of monster homes.

Columbia Falls has much in common with Libby, a logging and mill town in neighboring Lincoln County. Libby's residents live almost entirely in mobiles. Trailer parks cluster along the main thoroughfare near the center of town, and double-wides line gravel roads leading to logging areas outside town. Locals have taken to nailing green wooden signs to fences and mailbox posts declaring, "This Family Supported by the Timber Industry." Like other resource towns its size, Libby has suffered through the boom-and-bust cycles that follow logging activity. Manufactured housing provided a quick, affordable, and transportable form of housing during rapid expansion, and remained the only affordable home option during recession or downsizing.

Hungry Horse, Olney, Creston, and other small settlements throughout northwestern Montana are all changing in some way, be it learning to live with new residents or saying goodbye to the old. Towns closest to Glacier National Park, Big Mountain, and Flathead Lake cater to tourists as well. Those farther away from Kalispell or major tourism facilities continue to adjust to a shaky timber industry. Counties elsewhere in western Montana are also changing rapidly, and in the eastern half of the state, where teenagers leave for college and seldom come back, an aging but stable population creates a growing demand for health care services and affordable housing.

In 1990 more than 15 percent of Montana's housing stock consisted of mobile homes. Other western states, including New Mexico, Wyoming, and Arizona, had even more. Perhaps most telling about how mobile home owners live is that only 12 percent of all mobiles in the state are in parks. The other 88 percent are on forest roads and city streets, in rural valleys, farms, and amenity towns. For the past 20 years mobiles have remained the fastest-growing sector of the state and regional housing market, and for many the mobile represents their "dream home."

Montanans struggle to find adequate housing. Since World War II, and particularly in the past 30 years, mobile homes have provided the answer. Now, as the state continues to grow, the role of manufactured housing has become even more important. After several decades of little or negative population growth, the state's population jumped from 803,000 in 1990 to 902,195 in 2000, an increase of almost 13 percent. The Census Bureau predicts that Montana's population will reach 1.015 million by the year 2010.

The newest population surges makes an already difficult situation—the shortage of affordable housing—even more severe. The State Wide Housing Crisis Task Force has considered a number of options, and state agencies have responded to the shortage in part by providing low-interest loans to mobile home buyers who will be locating the home outside a trailer park.

Rural residents buy more mobile homes per capita than urban residents, and the smaller the city, the more mobiles it probably has. Even Billings and Great Falls, however, are home to several thousand mobiles, mostly in parks. Just outside the cities, the housing stock is comprised of mobiles, which constitute almost half of all housing in the urban fringe. Missoula and Mineral Counties present perhaps the most visible examples of how important mobile homes have become. During the 1980s, 79 percent of new homes in the Missoula urban area were manufactured. By 2000, Missoula County had 175 mobile home parks.

Throughout the state, with or without loans, high demand for attractive and affordable housing keeps mobile home sales brisk and the highways clogged with semitrucks hauling sections of double- and triple-wides. Dealerships straddle the major roads leading into the state's larger cities. In Missoula, Highway 10 between the airport and downtown is literally a neighborhood for sale—dozens of mobile home sales centers and the occasional trailer park in between. Several dealerships line Highway 2 in Evergreen, selling homes to Kalispell-area residents and potential homeowners from a three-county area.

The Changing "Small"-Town West

The small-town American West finds itself in transition at the turn of the century. Coping with continued population growth in an increasingly service-oriented economy, small cities and rural communities have had to ask themselves how big they want to become. The tide of exurbanites will continue to inundate amenity towns in the Intermountain West, and migrants from economically depressed areas, such as eastern Montana, will be lured to larger towns in the hope of finding work. The question, then, is,

How do communities prepare themselves for change and an increasingly service-oriented economy? How much sprawl is too much, and what kind of development do we want? Some answers lie in planning and business initiatives, but an antiregulatory western mentality limits the effectiveness of such initiatives.

Changes in the economy only strengthen the popularity of mobile homes. A job at the corner convenience store or at the Super 8 Motel pays less than mill work. The average income for a Flathead Valley resident now tops out at $6 an hour, according to the Flathead Job Service. Couple the low wages with rising housing costs within the city, and mobile homes make the most sense to potential homebuyers. Instead of purchasing a home near the center of town, an increasing number of residents are choosing to buy a large parcel of land in a rural area and to buy a less expensive double-wide. It's a logical choice in places where what is really for sale is a bit of breathing room. Both longtime and new residents are moving farther out. The former often find that the small town they once knew has changed too much for their liking. Folks new to town, many of whom are used to hour-long commutes from the Seattle suburbs to Boeing, think nothing of a 20-minute drive from a house ten miles or so outside town.

Mobile homes now account for one of every four new single-family homes sold in the United States, but these numbers are significantly higher in the South and West. The West has had a long time to get used to the idea of mobile homes, and the boom-bust nature of many of its resource towns has meant that residents have occasionally had few other options. Although the boom-bust cycle is less dramatic today than 20 or 30 years ago, exurbanization places many of the same demands on a town to provide the same amount of housing as a small-scale logging project or the opening of a new plant.

Mobile home dealership strips similar to those in Montana are to be found elsewhere in the West. Moses Lake, Washington, smack-dab in the middle of the state, has long served as a getaway for Puget Sounders interested in a drier climate. The resorts still line the lakeshore, but today it is mobile home, not boat, dealerships that claim the Business 90 strip. Southern Oregon, northeastern Wyoming, and urban fringes in the Southwest are the modern-day equivalent of a gold rush for mobile home dealers.

And why should it not be so? People have outdated images of mobile homes. The shoddy trailer of the 1960s is not the same product being sold on the market today. Welcome in the double-wide with stained glass and decorative trim, the triple-wide with the jacuzzi and cathedral ceilings. Thousands, if not tens of thousands, of 1960s- and 1970s-model mobile homes remain in the West, raised, altered, and repainted; they lasted longer even than dealers might have predicted three decades ago. For some 15 years now the manufactured-housing industry has been successfully producing a product that homebuyers find attractive, during a time when "bigger is better (and more expensive)" became the unspoken mantra of the site-built housing industry.

Manufactured housing no longer appeals solely to the low-income westerner. Middle-income buyers, frustrated at the lack of affordable starter and two-story homes, opt for mobiles as the next best thing. Mobiles meet federal housing and safety standards. Double- and triple-wides are indistinguishable from site-builts. Even sales of single-wides remain strong. Despite the tornado jokes and trailer-trash references, low- *and* middle-income westerners are finding that buying a mobile home just makes sense.

PART FOUR

PARKS

Mobile Home Parks, Utilitarian to Upscale

Virtually every place in the United States, no matter how small, has a cluster of mobile homes out at the edge of town. Larger places have more and larger clusters, many formalized as mobile home parks. Clusters of 20 or fewer units are generally fairly informal, with unpaved streets, few facilities, and haphazardly sited units. Parks with 50 or more units are generally planned and laid out more meticulously, with more order, more structure, more sense of a strong controlling hand. Significant economies of scale start to kick in somewhere between 30 and 100 units, which is the minimum size to warrant a full-time resident manager.

Mobile home parks vary enormously, and we need all kinds, from simple utilitarian parks with long harsh rows of gleaming metal boxes lined up with military precision to carefully landscaped upscale mobile home communities that would put most suburbs to shame. Their range of quality has widened enormously in recent years. The best have become very good indeed, while the poorest remain ominous and repugnant, and a few are so threatening that we feared to venture into them.

Most of the newest and best upscale mobile home communities are in well-to-do retirement areas in Florida, California, and Arizona. Utilitarian parks are everywhere, wherever low-income people need inexpensive (and quick) housing: on the fringes of cities and towns, in rural areas for migrant farmworkers, near construction sites and industrial plants, especially meatpacking plants, near military bases, and as emergency housing where disasters have decimated the housing stock. They have not strayed far from the

Fig. 44. Mobile home park southwest of Detroit, Michigan

Fig. 45. Utilitarian mobile home park with unpaved sandy streets west of Clermont, Florida

old original easy-in easy-out layout, with long straight lines of long rectangular boxes.

Mobile home parks are unusual residential areas in that the residents own their homes but do not own the land on which their homes are sited. They pay the park owner a monthly fee to cover lot rental, services, and facilities. They pay personal property taxes on their homes, but the park owner pays the real estate tax on the land. Most smaller parks are independently owned and managed, but larger parks may be owned by mobile home dealers, by manufacturers, by national chains, or by institutional investors.

In the mid-1990s, the managers of real estate investment trusts (REITs) decided that mobile home parks with 100 spaces or more were a better investment opportunity than apartment houses. Management and maintenance expenses were modest, rents were a regular and reliable source of income, and mobile home parks had low turnover rates because of the high cost of relocating units. It is hard for delinquent tenants to skip out from a mobile home park without paying their rent, as they can from an apartment, and the management can slap a lien on the units of tenants who are too delinquent. Mobile home parks were an especially attractive investment near cities, where land prices were so high that individuals could not afford to buy lots on which to site their units, and in the year 2000 it was estimated that the ten largest mobile home REITs controlled 2 percent of all mobile home parks. In 1996, ROC Communities, one of the larger REITs, owned and managed 110 parks with 28,000 rental sites.

Critics have enjoyed a field day cataloging the problems that residents have had with the owners of older utilitarian parks, especially in urban fringe areas. Inflation and rising taxes will inevitably force even the best-intentioned park owners to increase their lot rents over time, and retired people on low fixed incomes will just as inevitably complain bitterly about any rent increase, no matter how necessary, reasonable, and justified it may be.

Legally, mobile home parks are the private property of their owners, who are free to impose whatever rules and regulations they wish, and to change them whenever they see fit to do so. In the old days parks for transient trailers had strict rules to protect the other residents as well as the park owners, and that tradition persists in some contemporary parks, especially where park vacancies are scarce, as they are in many urban areas.

Few new utilitarian mobile home parks are being developed in urban areas that need the inexpensive housing they provide. Complying with byzantine zoning and building regulations may require months or even years of development time, and the cost of development may run $20,000 to $25,000 a site. Suitable land is hard to find because planners are reluctant to make provisions for mobile home parks, and they assiduously protect fringe areas for agriculture. Many potential developers have decided to invest elsewhere.

Most existing mobile home parks are fully occupied, with waiting lists. The difficulty of developing new parks, the shortage of rental sites, and the immobility of units have given park owners considerable license in setting rules, rents, fees, and other conditions of tenancy. Some tenants complain that they feel like helpless captives in mobile homes that are permanently fixed on sites they merely rent, sometimes on only a month-to-month basis. They say they have only limited legal recourse against greedy and unscrupulous park owners, who may victimize them with arbitrary rent increases, inadequate park maintenance, hidden fees, unreasonable rules and regulations, and "tie-ins."

A tie-in is an arrangement that requires all tenants to buy their homes from the park owner, who also owns a dealership. Many mobile homes are bought secondhand, but tie-ins prevent tenants from buying cheaper used units unless the owner happens to have one for sale. Park owners may also require tenants to buy goods and services, such as cable television, propane gas, insurance, and home improvements only from vendors who will give the owner a kickback.

Tie-ins also force departing tenants to sell their homes through the park owner, and most tenants do sell their homes when they leave a park. Because of state licensing fees and highway restrictions on oversized loads, it is usually cheaper to buy a new home at the new location than to pay thousands of dollars to move the old unit and run the risk of serious damage in transit. In 1990 the cost of relocation ran around $2,500 to $3,000 for a single-wide and $10,000 to $12,000 for a double-wide.

Closing a park may aggravate the existing shortage of park spaces by forcing tenants onto an already tight market. Many older urban parks were originally in outer fringe areas that urban growth has turned into prime areas for more intensive development. Some owners actually developed their parks as temporary uses on land that was zoned commercial, with the intention of converting them to more lucrative uses when the time was ripe.

The closing of a park can be especially difficult for low-income residents, many of whom live in older units that may not be able to stand the stress of transportation to another park. Other parks may refuse to accept older units, because they prefer new units that will upgrade the appearance of the park and perhaps even provide the park owner with a kickback on their sale.

Consumer advocates argue that an owner who plans to sell a park should give its residents the right of first refusal, the right to buy the land before it is sold to anyone else, although its fair market value is often more than they can afford. The advocates add that the owner should be required to give substantial notice of intent to sell in order to give the residents time to organize a purchase or to relocate, and that the owner who decides to sell should be required to buy all units at their fair market value or to pay the cost of relocating them to another park within a reasonable distance.

Some park owners have decided to solve their headaches by converting their parks into condominiums or cooperatives and selling them to the residents. Cooperatives may be less cumbersome than condominiums, because each resident of a cooperative owns a share of the entire park, whereas the residents of a condominium own particular lots, an arrangement that requires the expense of surveying and preparing legal deeds for each lot.

Visits to many parks suggest that the critics have identified some real problems, but they may have overstated their case, because most utilitarian parks seem to be reasonably happy places. A certain tension between owners and residents probably is inevitable, but the handful of owners who are unscrupulous and residents who are malcontent may have received more attention than they deserve.

Utilitarian mobile home parks perforce are semipublic places. Single-wides are so cramped that residents spend much of their lives outside, and lots are so small that parks have less privacy and far more social interaction than conventional residential suburbs. People are friendly and curious, and most parks seem to be genuinely caring communities. Life is in the narrow streets, which are lined with parked cars, many with hoods raised, seeming to devour the young men half-buried in their engines.

Many parks now have full-time resident managers. The managers of mobile home parks must be genial and gregarious, because they have to deal with the public on a 24-hour basis. (Managers of parks with rental units are the exception: they must be more suspicious, because some renters are irresponsible.) The best managers make a special effort to communicate with their residents, and they like to work with their park tenants' associations to maintain and improve the quality of their parks.

Utilitarian mobile home parks are truly remarkable pools of talent. Many of the residents are skilled artisans who enjoy doing things with their hands, such as maintaining and adding on to their units, and they are willing to help out on repair and maintenance jobs in the park. "Anything that goes wrong in the park can be fixed by somebody who lives here," we were told more than once.

Upscale mobile home communities are a relatively recent development. They are designed to be more attractive than utilitarian parks, most of which have become permanent residential areas but still retain the traditional cramped layout of parks for transient travel trailers, with small lots and few amenities. The people who bought expensive new mobile homes as permanent residences wanted nicer places to put them, and the new upscale parks are laid out like conventional residential subdivisions, from which they are distinguishable only by the nature of their housing stock. In turn, the availability of nice places to put them has encouraged more people to buy newer and more expensive models of mobile homes.

Handsome new mobile homes are especially attractive to families who wish to move from rental apartments, to empty-nesters who wish to sell their site-built houses and downsize, and to retired people, many of whom want affordable second homes in the Sunbelt. Most of the new upscale mobile home communities are in Sunbelt areas, although some have also been developed near major cities in other parts of the country.

The developers of the new upscale mobile home communities are targeting well-to-do consumers with an impressive range of amenities. Many focus on golf courses and marinas. Their clubhouses may have restaurants and sandwich shops, post offices and convenience stores, card and billiard rooms, dance floors, and saunas and jacuzzis in the exercise room. Next to the clubhouse may be the heated swimming pool, tennis and shuffleboard courts, putting greens and bowling greens, croquet courts, a horseshoe pitch, and perhaps even a softball diamond.

Many of the new parks have guarded gatehouses with 24-hour security, which senior citizens especially appreciate. The curving streets are paved and curbed, with distinctive lights and signs, and the park maintains the lawns meticulously. The lots are large enough that each home has a two-car parking apron and a carport on one side, and a patio with privacy fencing on the other. The cars parked in the driveways and the boats tied up in the canal out back testify that the residents are people of ample means.

Many of the residents of the new upscale parks in the Sunbelt are "snowbirds" who simply lock up the place in the spring and drive to their summer

Fig. 46. Mobile home park under the palms east of Okeechobee, Florida

Fig. 47. Wharves for pleasure boats in the canal behind mobile homes in Trailer Estates, Florida

homes up north. They came initially to escape the northern winter, but in the park they have developed friendships that keep drawing them back. They have large amounts of discretionary time, but the park keeps them fully occupied with a frenzied social whirl of club meetings, dances, classes, tournaments, potluck meals, and similar group activities.

Each utilitarian mobile home park and each upscale mobile home community has its own distinctive history, and its own demographic and socioeconomic profile. We have sampled a number of individual parks to give something of the flavor of their uniqueness and diversity.

The Twin Cities Metropolitan Area

The Twin Cities (Minneapolis and St. Paul) metropolitan area in Minnesota illustrates the complex distribution of mobile homes in a metropolitan area. This metropolitan area has been defined four times. In 1966 the state legislature created a Metropolitan Council to direct urban growth in a seven-county area by controlling investment in public facilities and services, but the Metro Council area is not congruent with any of the metropolitan areas defined by the U.S. Bureau of the Census in 1949, in 1972, and in 1983.

The Metro Council area is a political crazy quilt. Minnesotans do not like the idea of cities annexing adjacent areas, but they also dislike telling other people what they can or cannot do, so in typical Minnesota fashion they have tried to make annexation harder by making defensive incorporation easier. As a consequence, the Metro Council area includes 7 counties, 9 metropolitan agencies, 22 special districts, 49 school districts, 50 independent townships, and 138 incorporated municipalities.

Each polity has its own ideas about mobile homes. About half do their best to discourage them, most of the rest permit them on a permanent basis only in parks, a few permit single-sited units, and two (Hilltop and Landfall) have actually incorporated to preserve their identity as mobile home communities.

In 1998 the Metro Council area had 89 mobile home parks, ranging from 16 to 570 spaces, with an average of 175 spaces per park. These numbers have hardly changed in two decades, and park space is in chronically short supply. A few parks have been closed down, and a few new parks have opened, but increasing land prices and the difficulty of finding sites where parks are allowed have discouraged the development of new parks.

Most mobile home parks in the Twin Cities area are on the outer fringes of the built-up area, although older parks that were once on the fringe have been engulfed by subsequent urban growth. The heaviest concentrations of parks are in the low-income, blue-collar suburbs north of the two central cities (Fig. 48). In 1998 the city of Blaine had six mobile home parks with a total of 2,366 spaces.

On April 28, 1996, the *Minneapolis Star Tribune* reinforced a popular image by reporting that 31 percent of Blaine's 1,161 police calls for domestic violence came from mobile home parks, although only 18 percent of the city's housing units were mobile homes. "It's a densely populated area," said a city police supervisor. "You've got a different income level here, and some of the mentality is different. People of moderate income are more likely to call the police, while people with money try to hush it up. And," he added pensively, "the police are more likely to intervene in the lives of poor people."

Hilltop, which is just across the county line north of Minneapolis, and Landfall, which is just across the county line east of St. Paul, are the two municipalities that incorporated to protect their identity as mobile home communities. Hilltop is at the southern tip of blue-collar Anoka County, which is wedged in between Hennepin (Minneapolis) and Ramsey (St. Paul) Counties (Fig. 49).

Hilltop began as a 40-space trailer park in the open country in southern Anoka's Fridley Township, which later incorporated as the city of Fridley. The park was next to two-lane State Highway 65. A successful magazine salesman named Leslie Johnson, who had to travel a lot, had a flat tire here one day in the late 1940s. He stopped at the trailer park to fix it and made instant friends, so he returned frequently. His wife was fed up with traveling, and when she heard that the owner was planning to sell the park, she told her husband that they were going to buy it.

Johnson bought the park in 1953 and quickly developed it into Trailer City, with 164 spaces, and in 1955 Ervin Shear developed 31-space Sunnyside Trailer Park north of it. Fridley Township was growing explosively in the early 1950s. The city of Columbia Heights was in an expansionist mode, aggressively annexing adjacent areas, and it decided to annex these two trailer parks to get rid of them.

Les Johnson fought back. He led a drive to incorporate an 80-acre rectangle, four blocks wide and five blocks long, to protect the trailer parks it contained. This area had 688 residents in 240 homes, of which 195 were trailers. On May 1, 1956, a total of 171 residents trooped to the polls, and 137 voted to incorporate Hilltop. The village took its name from the Hilltop drive-in theater, which was on the site where an old dairy farm had been developed into the Hilltop Stables and Riding Academy.

Hilltop has subsequently added two more mobile home parks (with a total of 92 spaces), and several apartment complexes (with a total of 69 units) were built in the 1960s. Central Avenue has been expanded into a major four-lane arterial thoroughfare, and it has become a busy commercial, apartment, and office strip (Fig. 50).

Fig. 48. Mobile homes in the greater Minneapolis–St. Paul area, 1990

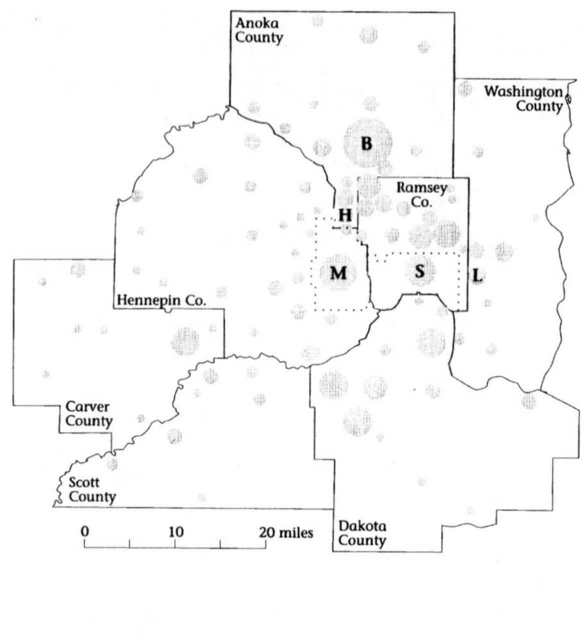

Fig. 49. The city of Hilltop, Minnesota, is completely enclosed by the city of Columbia Heights at the southern tip of Anoka County between Hennepin and Ramsey Counties

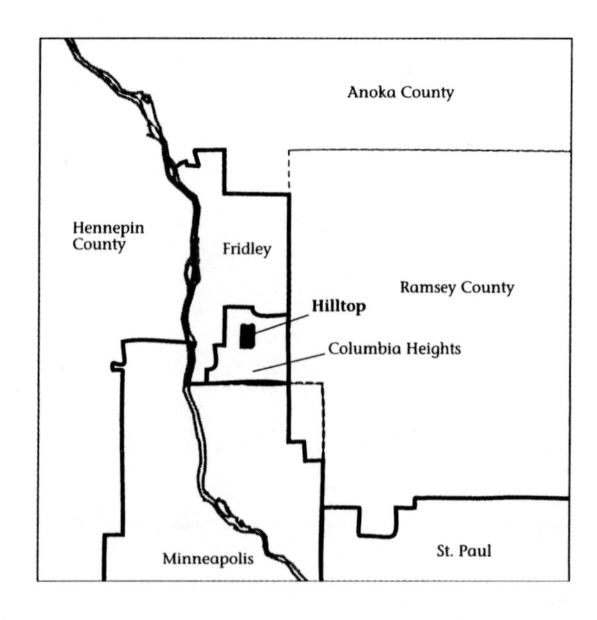

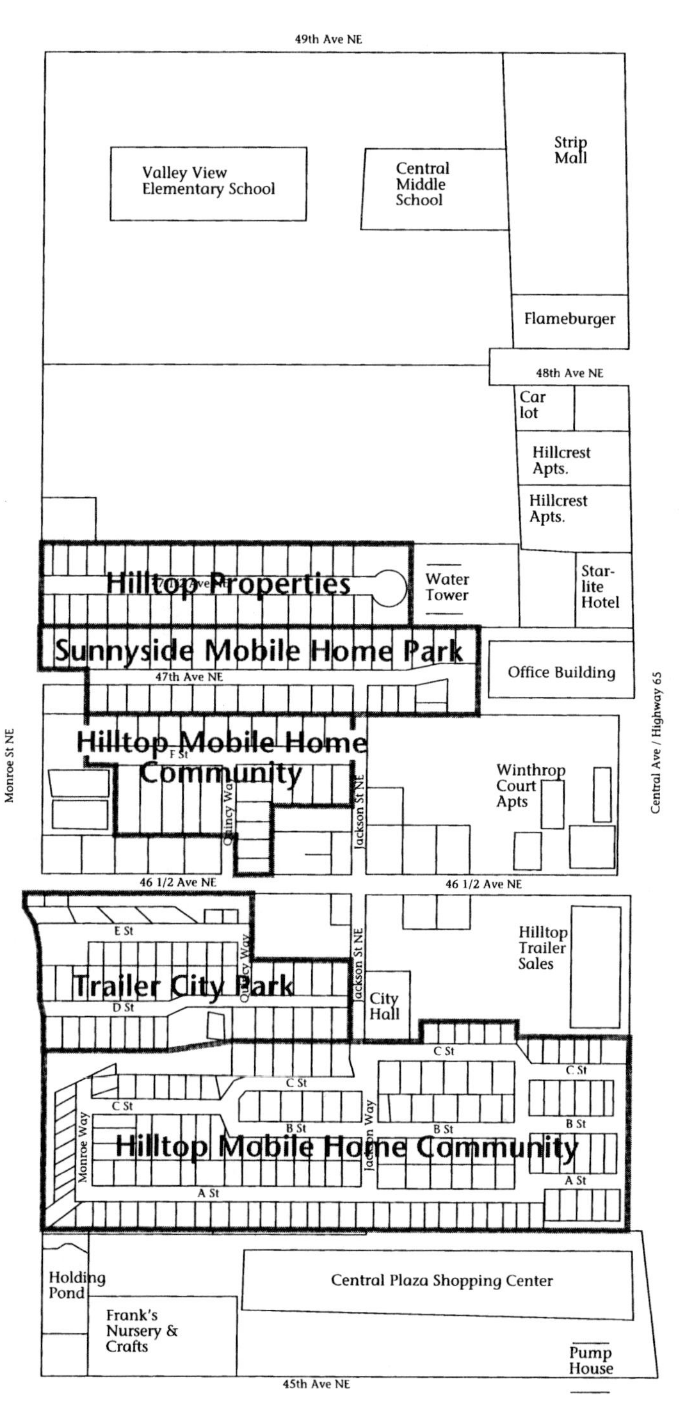

Fig. 50. Land use in Hilltop, Minnesota, 1996

Hilltop has been a perennial source of irritation to Columbia Heights, although things have calmed down considerably since the disputatious early days, which were marked by constant squabbles and takeover efforts as Columbia Heights gradually surrounded Hilltop. Columbia Heights struck first in 1957 by cutting off Hilltop's water and sewer services, and Hilltop sold bonds to develop its own water, sanitary, and storm sewer systems.

Columbia Heights was irate when Hilltop permitted private liquor stores that drew business from the municipal liquor store, whose profits accounted for one-third of Columbia Heights' total operating budget. The conservative merchants of Columbia Heights blocked a developer who wanted to build a shopping center, but Hilltop welcomed him, and he opened the Central Plaza Shopping Center on the southern ten acres of Hilltop in 1958. Columbia Heights retaliated by refusing to let him hook on to its sewer system, but later relented.

Hilltop had to shop around for other municipal services. Schools were no problem, because both Hilltop and Columbia Heights are in School District 13, which owns ten acres in northern Hilltop, where Valley View Elementary School was opened in 1960 and Central Junior High School in 1965. In 1956 Hilltop contracted with Fridley for fire protection, but in 1964 Fridley refused to renew this contract, and Hilltop had to contract with Spring Lake Park, seven long miles away, until Columbia Heights relented and agreed to a contract in 1968.

In the early days Hilltop had its own police department, with a full-time chief and four part-time helpers. The chief once had to fire one overly zealous part-timer because he had given the village a reputation as a speed trap when he took it upon himself to start ticketing cars on Central Avenue. In 1961 the chief asked the city council to buy him a gun powerful enough to shoot through a car's engine block, but the council denied his request when the city attorney said, "You shoot that thing here and you'll go through six trailers before you'll hit a car." The police department was closed in 1972 when the chief wrecked the city's one and only squad car. The council decided that it could not afford to buy a new one, and it contracted with Columbia Heights for police protection. "No big deal," said the mayor. "A police officer taking his time can patrol all of Hilltop in 30 minutes."

In 1999 Hilltop bought its water from Minneapolis and contracted fire protection from Fridley and police and sewer services from Columbia

Heights, but Hilltoppers are proud that they have always had a hands-on kind of community where people pitch in and do what has to be done. Back in 1963 Mayor Bruce Hay had told a reporter, "When the sewers are plugged, we get out a hose and flush them out. When street signs have to be put up, the trustees and I go out and put them up. When the streets are slippery, some volunteers borrow a truck, pick up some sand, and start sanding."

Hilltop suffered a serious financial crisis in 1991. The city (a 1974 state law classifies all incorporated places in Minnesota as cities) was having trouble paying its bills on time and seemed close to bankruptcy. An investigation revealed that the city clerk, the only full-time city employee, who had worked for the city for 20 years, had embezzled more than $200,000. Ruth Nelson, who was hired to replace her as city clerk, has gotten Hilltop back on a firm financial footing. Her office is in the new stone city hall, which was built in 1987 to replace an older one that had burned. In 1998 she administered a budget of $475,000, half from commercial properties, one-quarter from mobile home parks, one-fifth from apartment buildings, and one-twentieth from conventional residences.

In 1999 Hilltop secured a grant from the Minnesota Housing and Finance Agency to support a manufactured-home renewal program. This program provides purchase assistance to enable owners to replace their older homes with newer models, and down-payment assistance to help first-time owners buy new homes. The city was also developing a proposal for state funding to hire professionals to help it renovate and modernize its utility system.

Contemporary Hilltop is dominated by the 135-foot-high spheroidal water tank the city built in 1964 to maintain a constant pressure of 50 pounds. It is painted bright blue, with "HILLTOP" in bold white block letters on either side, and it is festooned with communications antennas and warning lights. The water tank is a landmark, but the Central Avenue strip cordons off the mobile home parks from public gaze, and passing motorists might not even realize that they exist.

The principal clue on Central Avenue to Hilltop's mobile home parks has been the office and sales lot of Hilltop Trailer Sales. Originally this business sold house trailers, but it shifted to travel trailers when the parks in Hilltop filled up, and in 1999 it decided to pack up and leave because it needed more space.

Linda Johnson, Les Johnson's daughter-in-law, manages Trailer City Mobile Home Park. She has lived here 30 years, her sister-in-law lives next door, and her son lives across the street. "This park is not a way station," she said. "A couple of spaces seem to turn over every two or three years, but most people stay in my park a long time. The majority have been here 15 to 20 years. The oldest home is from around 1956, and it is still in good shape. We never have a vacant space, but people are calling me all the time asking about available space. This is a desirable location because we are close to the city, we are on bus lines, and people can walk to everything they need in the commercial areas."

Linda said that 95 percent of the residents work, and none of them are on welfare, because welfare won't pay the bills to live here. Spaces in the park rent for around $250 per month, depending on their size and location. Taxes vary with age and size, ranging from $150 a year for the older singles up to $325 a year for newer double-wides. The average value of homes is $10,000 to $15,000, up to $40,000 for new double-wides, as low as $5,000 for older singles. The park has no rental units, Linda said, because renters do not take the same pride in maintenance as owners.

"Everybody would prefer a house," Linda said, "but some choose to downsize. Some people have sold their houses and moved here so they can lock up and leave for the winter, or so they can afford a summer home at the lake. A mobile home is better than an apartment because you have identity, space, a garden, trees, a sense of ownership. It's a great starter home for a young couple, and it is really a nice, convenient home for a single person."

All utilities in the park—telephone, electric, gas, oil, water, sewer—are underground, the lots are sodded and landscaped, and the streets have lantern-type light posts. Most units are single-wides, but the trees that shade them soften the harsh angularity of the long serried rows. Steve Johnson, Linda's husband, shares his father's love for trees. He likes to plant them from seed, and he says that one tree per yard is enough. "I plant three," he said, "and then cut down two when one gets established. We lost 300 elm trees, and I have tried various types, but now I am mainly into hardwoods, even oaks."

The tiny city of Hilltop has offered a tempting target for humorists and would-be humorists on the metropolitan daily newspapers, and Hilltoppers have learned to take their barbs in stride, but they have a strong sense of pride in their community. Perhaps it is best expressed by the story of the five-year-old who came home from his first day at kindergarten and told his mother

Fig. 51. A double-wide mobile home in Hilltop, Minnesota

Fig. 52. Steve Johnson shows off a single-wide mobile home in Trailer City Mobile Home Park, Hilltop, Minnesota

Fig. 53. Different kinds of mobile homes line a tree-shaded street in Trailer City Mobile Home Park, Hilltop, Minnesota

that the kid seated next to him lived in the United States of America. "So do you," she said. "No I don't," he expostulated. "I live in Hilltop!"

The residents of Landfall, the only other incorporated mobile home community in the Twin Cities metropolitan area, have an equally strong sense of pride. Landfall was created by James E. Olson, who was 21 years old when he graduated from Macalester College in 1951 with a degree in theater. He married Mitzi Marie Hunter, and they set off in a travel trailer for advanced study at the University of Denver. The next summer they drove east in the trailer to work in summer-stock theaters. Olson quickly decided that the only people who were making any money in the theater business were the owners, and he made up his mind to own something.

The Olsons had stayed at some unattractive trailer parks in their travels, and he decided to develop one that would be a real showplace. In 1953 he bought 53 acres southeast of Tanner's Lake on U.S. 12, the principal highway into the Twin Cities from the east, which has since been upgraded to Interstate 94. The site was just across the Ramsey County line in Washington County's Oakdale Township, which was still completely rural. The site even had an abandoned dairy barn and silo when Olson bought it.

Olson named his park Landfall, after his favorite wharfside restaurant in Massachusetts, across the bay from Martha's Vineyard. ("Landfall" is an old sailor's term for the first welcome sight of land after a long sea voyage.) He developed the park in four stages. The first 20 spaces were in the clump of trees next to the lake, and this area also had spaces to rent overnight to transient travel trailers. In 1964 Olson opened 60 more spaces on the hilltop, in 1965 he developed 165 spaces north along the lake and on the terrace that overlooks it, and in 1971 he added 70 spaces back of the commercial strip. In 1992 the original part of the park was redeveloped into larger spaces for double-wides, which form a handsome gateway to the community.

The commercial strip along the frontage road has had several incarnations. In 1953 Olson leased land along the highway to the Pure Oil Company for a truck stop. In 1973 he converted the Pure Oil building into a dinner theater, which was followed by a series of other equally unsuccessful ventures until 1997, when the property was sold to the used car business and auto repair service that occupied it in 1999. In 1969 Olson sold the southwestern corner of the village to a home and garden center, which in turn sold it to a motorcycle dealership in 1997.

Fig. 54. Landfall, Minnesota, in August 1960, before the upper level in the upper right corner had been developed. View from south of U.S. 12, which was subsequently upgraded to I-94. Photograph courtesy of Helen Hallis.

Fig. 55. Birch Lane in Landfall, Minnesota, has been redeveloped with lots for double-wide mobile homes

Landfall was in Oakdale Township, which was unabashedly rural, and the township board was singularly insensitive to its needs. When the board thought about the park at all, which was not often, it merely wished that Landfall would go away, because it was not kindly disposed toward trailers or trailer parks. James Olson convinced his residents that incorporation was the only way to obtain the municipal services that the township refused to provide.

In Minnesota the only legal requirement for a vote to incorporate was a petition signed by 100 residents, and Landfall had 313, so on April 6, 1959, Landfall incorporated by a vote of 99 to 2. The incorporation of Landfall produced the extraordinary situation of a municipality whose entire land area was owned by a single individual, but the city and its owner seem to have worked out a mutually agreeable allocation of responsibilities. The city contracted with other municipalities to provide police and fire protection, and the owner was responsible for providing water, sewers, and utilities, and for maintaining streets and sidewalks.

James Olson died in 1985, but even before his death he seems to have lost interest in Landfall. "He liked to own land," said Helen Hallis, the city clerk. "He owned two other mobile home parks, and he had real estate interests all over the place, but he forgot about Landfall. The park was really going downhill, because the Olsons were not putting any money back into it, and they allowed it to run down for lack of proper maintenance."

In 1991 James Olson's widow, Mitzi, decided to sell Landfall, and she put it on the market with a price tag of $6.3 million. Landfall is prime real estate, because it is next to an interstate interchange and virtually in the shadow of the elegant headquarters campus of the Minnesota Mining and Manufacturing (3M) Company. The residents were concerned that some private developer would buy the property, evict them, and build a shopping mall or a luxury complex, so they decided that the city should buy it to give them security.

The city of Landfall had no bonding capability, so it appealed to the Washington County Housing and Redevelopment Authority (HRA) to help it buy itself. The county HRA is charged with providing low- and moderate-income housing that is safe and decent, and in order to preserve 300 units of low-income housing in Landfall, which was rated the most affordable suburb in the metropolitan area, it agreed to sell the necessary bonds, to be repaid by tax levies over a period of ten years.

The Washington County HRA instructed the city of Landfall to create its own HRA to operate the park—which was fortunate, because in 1997 some severe financial problems forced the Washington County HRA to retrench. Landfall had to refinance its bond in order to enable the Landfall HRA to buy the city from the Washington County HRA for $7.2 million, which it expects to pay off in 20 years.

As if this were not confusing enough, the Landfall HRA and the city of Landfall are separate entities, but the members of the city council also form the board of the HRA; on Monday nights they wear their city council hats, and on Tuesday nights they wear their HRA board hats. Residents pay property taxes on their homes to the county, and they pay rent for their spaces to the HRA, which collects around $1 million each year. The HRA must pay off the bonds and pay county real estate taxes and other expenses of ownership.

The city provides garbage, sewer, and water services, maintains the streets, and pays the city of Maplewood $120,000 a year for police protection and $17,000 a year for fire protection. The city has an annual budget of $400,000, of which about 15 percent comes from taxes on commercial property, and most of the rest from a state "fiscal disparities" fund.

In 1971 Minnesota passed a fiscal disparities act to lessen differences in the tax bases of Metro Council communities. Each community contributes 40 percent of its commercial-industrial tax base growth since 1971 to a regional pool, and then receives back a share of the pool in proportion to its population and tax base. Municipalities with good tax bases per capita receive less than they contribute, while those with poor tax bases, like Landfall, receive more.

Landfall City Clerk Helen Hallis oversees a staff of seven full-time employees. "My goal," she said, "is to get Landfall back into being a showplace. We started in 1993, when we got a state grant for improvements in the water system, for housing rehabilitation, and for the construction of a community center and storm shelter. We need a community center in a place where the houses are so small that there is no place for groups to get together, and a storm shelter is essential in a community without basements. In 1996 we built a new $400,000 city hall with a nice view over Tanner's Lake."

Landfall has regularly surveyed its residents, and their responses indicate a satisfactory level of satisfaction, with a few predictable old grouches. The most recent survey, in 1995, revealed that the average resident had lived in Landfall eight years. Forty percent were aged 34 or younger, and 25 percent were 55 or older. Half had incomes of $10,000 to $25,000 a year, 14 percent made less than $10,000, 19 percent made between $25,000 and $35,000, and 15 percent made more than $35,000.

In 1995 Landfall had 37 lots for double-wides and 265 lots for single-wides, but the lots were so small that they could not handle units larger than 14 by 70 feet. Most single-wides were 15 to 25 years old; they generally sold for $5,000 to $20,000, with newer models selling for up to $30,000. Most double-wides were five to eight years old, and sold for $20,000 to $50,000. Lot rents ran from $270 to $290 a month for single-wides and $315 to $335 for double-wides, with higher rents near the lake. County property taxes ranged from $50 a year for old $5,000 single-wides to $200 a year for new double-wides.

Homes are rarely moved, although occasionally a resident will remove an older home and replace it with a newer model. A family that leaves will sell its home to a new family eager to move in. All homes are owner-occupied, and most are in good condition. In 1998 the city gave one month's rent free to those who painted the exteriors of their homes, which did a lot to spruce up the community. Fifteen percent of the homes need minor repairs or maintenance, and residents with low or fixed incomes may need financial assistance to do the necessary work.

Hilltop and Landfall both defy the conventional negative stereotype of mobile home parks, and they probably are more representative than the stereotype. Most residents take pride in their homes and in their communities, and they are continually upgrading and refurbishing both. Pride of ownership and pride in place are truly important.

Southwestern Kansas

At the eastern edge of Garden City, Kansas, a strip of low bushy austrees north of U.S. 50 tries to screen one of the largest mobile home parks in the United States. This park has 600 lots, and the owner plans to expand to 668. The lots, each measuring 50 by 100 feet, are large enough to handle double-wide units easily, but only a handful of the units are double-wides. Only ten lots are empty, and the park has a waiting list, even though it is grimly utilitarian, with units rigidly aligned in long, monotonous, parallel rows of 37. This park is merely the largest of many that serve the newly emergent meat-packing complex of southwestern Kansas.

Since 1960 southwestern Kansas has become one of the nation's leading beef-producing areas. In the 1880s local farmers began to divert water from the Arkansas River to irrigate the level bottomlands along the stream, and they grew sugar beets as their principal cash crop. Farmers on the dry uplands grew winter wheat, and they ran herds of beef cattle on the rangeland that was not suited to cultivation. In the 1950s sugar beets were becoming less profitable, and some local entrepreneurs developed feedlots, where they could bring lean cattle from the range and fatten them to market weight.

A few farmers pumped from shallow wells, but they irrigated most fields only by gravity until the 1960s, when the development of powerful turbine pumps enabled them to pull water from the famous Ogallala Aquifer, 200 feet below the surface. The development of center-pivot sprinkler irrigation systems enabled them to irrigate the rolling uplands, where they grew bumper crops of corn and alfalfa. The abundant supply of feed in the area encouraged the development of huge feedlots for beef cattle, and by 1990 southwestern Kansas had nearly 100 feedlots with a total capacity of more than 1 million beef cattle.

The availability of fed cattle, plus an abundance of water from the Ogallala Aquifer, attracted meat-packing companies to the area. In 1980, IBP, Inc. (formerly Iowa Beef Processors), opened the world's largest beef-packing plant in Holcomb, ten miles west of Garden City. The Holcomb plant employs 2,800 workers, and in 1983 Monfort modernized its Garden City plant and doubled its work force to 1,300. These two plants slaughter 8,400 cattle a day. In addition, Dodge City has two large beef-packing plants, Liberal has one, and Guymon, just across the state line in Oklahoma, has a major hog-slaughtering facility.

The new plants in southwestern Kansas highlight how dramatically the meat-packing industry has changed since World War II. It has moved from metropolitan areas to smaller places that are closer to the source of the animals, and it has replaced skilled union butchers with lower-wage, unskilled workers who constantly perform the same, simple, monotonous tasks on production lines. Carcasses or boxes roll inexorably past them at a rate of around 400 an hour, or one every nine seconds for the entire working day, and they must do precisely the same thing to each one.

Work in a meat-packing plant has always been brutish, it is poorly paid, and it is extremely hazardous. Meat packing is second only to underground mining as the nation's most dangerous industry. The knives and saws that slice up the animals are equally ready to butcher human flesh. The packing companies have been criticized so harshly for so long that they have become skittish, and they are understandably reluctant to answer any questions from outsiders. One company executive even refused to tell us the number of cars that could be parked in his company parking lot.

The meat-packing companies had to recruit new workers for their plants, because southwestern Kansas is sparsely populated, and few local people

Fig. 56. East Garden Village Mobile Home Park in Garden City, Kansas, has 16 straight rows of 37 mobile homes each. The cars parked in nearly every driveway even during normal working hours suggest that most families have at least two cars.

Fig. 57. The sign on the ice cream vendor's cart in the East Garden Village Mobile Home Park hints at the ethnic composition of the population of the park

were willing to take such dangerous jobs at such low wages. The companies recruited so well that the population of Garden City town and township boomed from 20,980 people in 1980 to 29,565 in 1990, with most of the growth in the first half of the decade, when Monfort doubled its work force. Some of the new people were Anglos, some were immigrants from Mexico, and some were refugees from Southeast Asia, mostly Vietnamese and Laotians.

Many of the newcomers are isolated from the local residents by language and customs as well as by occupation. The population of Garden City, for example, is about one-quarter Hispanic, but the "immigrants" who came directly from Mexico have had little contact with the "native" Hispanics, the established second- and third-generation descendants of those who came in the early 1900s to work in the sugar beet fields and on the railroads.

The overcrowding of housing units is legendary. Local folklore tells of a trailer park manager who investigated one unit that seemed to be using excessive amounts of electricity. He found that the occupant had cut a hole in the floor, and six Vietnamese men were sleeping on the ground beneath it. The story might be apocryphal, but it illustrates popular perceptions of trailer-park living and perhaps of the Vietnamese lifestyle.

Turnover is heavy. Half of the newcomers leave the area within a year, and two-thirds stay less than two years, so the packing companies must constantly be recruiting new workers. Recruitment from Mexico is largely by word of mouth, but we were told that officials in California subsidize recruiters from the packing companies to help reduce their welfare rolls by hauling busloads of Southeast Asian refugees to Kansas.

Minh Duong, personnel director of the IBP plant in Holcomb, said, "We started construction in 1979 and began processing in 1981, but this area had no housing for workers. Some of them lived in parking lots, and some camped out in parks. We had to delay starting a second shift until housing was available in a large trailer park east of Garden City. It is ten miles from our plant, but the land there was cheap. I have heard that they bought an old dump and leveled it."

"IBP encouraged contractors to develop trailer parks," said Duong, "by guaranteeing them referrals of our employees. At one time we deducted their rent payments directly from their paychecks, but that sort of died out around 1990 because we had enough housing. Most of our employees rent-to-own their units. The average price of a unit is around $14,000, and they rent a space in a park for $135 a month. Garden City has four trailer parks, and it doesn't want any more, so now IBP is trying to encourage contractors to build apartments that our employees can afford."

Rob Martin owns and operates the huge East Garden Village Mobile Home Park east of Garden City. "My parents have been in the 'wobbly box' business for many years," he said. They gradually built up a 520-unit park in Sioux City, Iowa, which is right across the Missouri River from IBP headquarters in Dakota City, Nebraska. When IBP came to Garden City, which needed housing badly, its managers asked Rob's father to co-venture with them. IBP put up 80 percent and promised to buy 500 units in increments of 125. Then the company was bought out, and IBP cut the program.

Rob has a degree in mechanical engineering from the University of Oklahoma. In 1984 his father sent him to Garden City to straighten things out. He built the park up to 280 units by using a variety of incentives. He paid the cost of moving a unit from anywhere within a 50-mile radius. He offered cash incentives and allowed lots to be occupied rent-free for the first month. He gave a month's free rent to anyone who could get a friend to move in, but he does not need to do that anymore. In 1986 he bought the 125 IBP units. He rented them until 1992, because he needed the cash flow. He had the park full by 1993 and plans to expand to 688 lots. He is going to develop a retail complex near the highway with a convenience store, a liquor store, and a car wash, and will have space to rent.

"We put the Garden City park where it is," he said, "because we had to be close to city utilities, we needed flat land, and we wanted space to expand. That story about the dump is a good story, but it's just a story. The park really is a white elephant, because it has gotten too big. Two hundred and fifty units is about the right size for a manageable park. The first hundred units should pay the overhead, and you can make money from the rest."

Rob is developing a second large mobile home park near Guymon, Oklahoma, where the Seaboard Corporation is developing a huge new hog production complex. He said,

> They have built a $100 million plant in Guymon that will process 4 million hogs a year, and they are spending $300 million to develop their own hog farms on land they have bought in a 60-mile radius. They plan to produce 2 million hogs a year from their own farms, and to buy 2 million more from other operators. The city failed to correct the local housing deficiency after they had built their processing plant, and they turned to me. They loaned me the money to buy the land and to make improvements, because I am stuck with it if they bail out, but I had to finance the units.
>
> I put in 101 units in 100 days. I bought 14-by-70-foot units for $18,000. They have three bedrooms and two baths. The rent is $400 a month with basic cable for rental units, or $125 a month just for the lot. Each Thursday, which is payday, Seaboard takes out a payroll deduction of $92.31, 12 months' rent divided by 52 weeks, but I don't see it 'til Monday. We also rent furniture, because we rent the units unfurnished. I expect to expand to 244 units. I want to keep around 100 rental units and plan to sell the rest for $20,000 to tenants who are

creditworthy. As soon as I sell 25 units I will build another street of 30 until I have developed a 244-unit park.

The Country View Mobile Home Park southeast of Guymon has the austere functional beauty of a place that is precisely what it says it is: an oasis of affordable housing for low-income people. At first glance it looks pretty bleak, with long straight rows of identical masonite boxes set on bare weed-speckled ground with not a tree or even a bush in sight. "I didn't have time for any landscaping," Rob said, "because Seaboard wanted 'em here in a hurry, and I had to do 'em quick. I am going to irrigate the whole park, and then I am going to put in sidewalks. I have tried to plant trees many times in the Garden City park, but the kids kill them as fast as I can plant them."

In Guymon, Rob is also developing Country Estates Mobile Home Park, with 100 lots for double-wide units. "The workers can afford them," he said, "but they can't afford the houses that some contractors are building. I am often asked if I am interested in developing new parks, and I'm still in the market if I can find a good opportunity. I need an outlying area close to city utilities, flat, and with room enough to expand. Seaboard wants me to put 100 units in Ulysses, Kansas, where they have a lot of hog farms. They want me to put in smaller parks in smaller places for their farmworkers."

Rob has parks in Garden City and in Guymon, and he spends one day a week at each one, but his family has moved to Colorado Springs because the schools are so much better. "I have my own plane," he said, "but this commuting gets old after a while. My mom and dad have sold our park in Sioux City, and they have 2,200 lots in the Lower Rio Grande Valley. Eventually I will probably sell the park in Garden City, because mobile home parks are a hot investment item right now. What will I do with the money? I might just put it in CDs, you never know."

Rob's plans were changed dramatically in the fall of 1995, when Seaboard decided to go to a second shift that would employ 800 to 900 more people in the hog-processing plant in Guymon. They already had workers driving in from 50 miles or more, and they realized that they needed more housing, so once again they turned to the old pro. They asked Rob to expand his park to 680 units, which would cost around $17 million. He had to borrow the money, and Seaboard satisfied his lenders with a lukewarm guarantee that they expected the plant to stay open for at least five years.

Fig. 58. Rob Martin has developed the Country View Mobile Home Park southeast of Guymon, Oklahoma, to house workers in a new hog-processing plant

Fig. 59. Rob Martin has not yet had time to landscape the Country View Mobile Home Park

All of the units will be rental units, and they will all belong to Rob after he has paid off his loan. "Managing rental units is at least five times the work of managing a park in which the tenants own their units," he said, "and it is going to be a real challenge, but I am looking forward to it. I am going to divide the park into blocks of about 125 units, and I am going to give one person full responsibility for each block."

Rapid industrial growth and rapid population growth often create a need for quick and inexpensive housing. Rob Martin is the largest developer of mobile home parks in Garden City and Guymon, and he has done a good job, but he is not an anomaly. Massive mobile home settlement and park development have accompanied the rapid growth of industry in rural areas throughout the South and West.

Long Neck

For more than a century the Atlantic coast of Delaware has been a popular playground for the cities of southern Megalopolis, the great blob of contiguous built-up area that stretches 400 miles from Washington northeast through Baltimore, Philadelphia, and New York to Boston. Each summer 3–4 million visitors descend upon the 25-mile strip of sandy beach between Cape Henlopen and Fenwick Island. The narrow strip of sand is chock-a-block with motels, eating places, shopping malls, and every other kind of establishment that can separate tourists from their money. Summer traffic jams are the stuff of legend, but still many visitors return year after year, and the fortunate few who can afford the astronomical prices have bought one of the summer homes that pack the beachfront like sardines.

Peninsulas are called "necks" in the Chesapeake country. Long Neck is the five-mile tongue of sand that separates Indian River Bay to the south from Rehoboth Bay to the north (Fig. 60). The highest point on Long Neck is all of 20 feet above sea level, and the lowest point is determined by the ebb and flow of the tides in the low-lying salt-grass marshes that rim the peninsula. The higher ground has fields of corn and soybeans and extensive hardwood forests.

In recent years resort development has been spreading inland in pursuit of cheaper land, lesser congestion, and greater peace and quiet. State Highway 1, the principal route toward Washington, Baltimore, and Philadelphia, has a number of small mobile home parks, but mobile home parks threaten to capsize Long Neck. Sussex County, Delaware, has more mobile homes per square mile than any other nonmetropolitan county in the United States. The U.S. Bureau of the Census recognizes that a new town has developed here, and in 1990 it identified Long Neck as a Census Designated Place (CDP) in which it reported that 87 percent of the housing units were mobile homes.

The most impressive mobile home parks on Long Neck are the seven Tunnell Communities, including four with the curious name Pot-Nets, which are more attractive than most suburban subdivisions. Their units are on unusually spacious lots, their lawns are groomed like golf course fairways, and

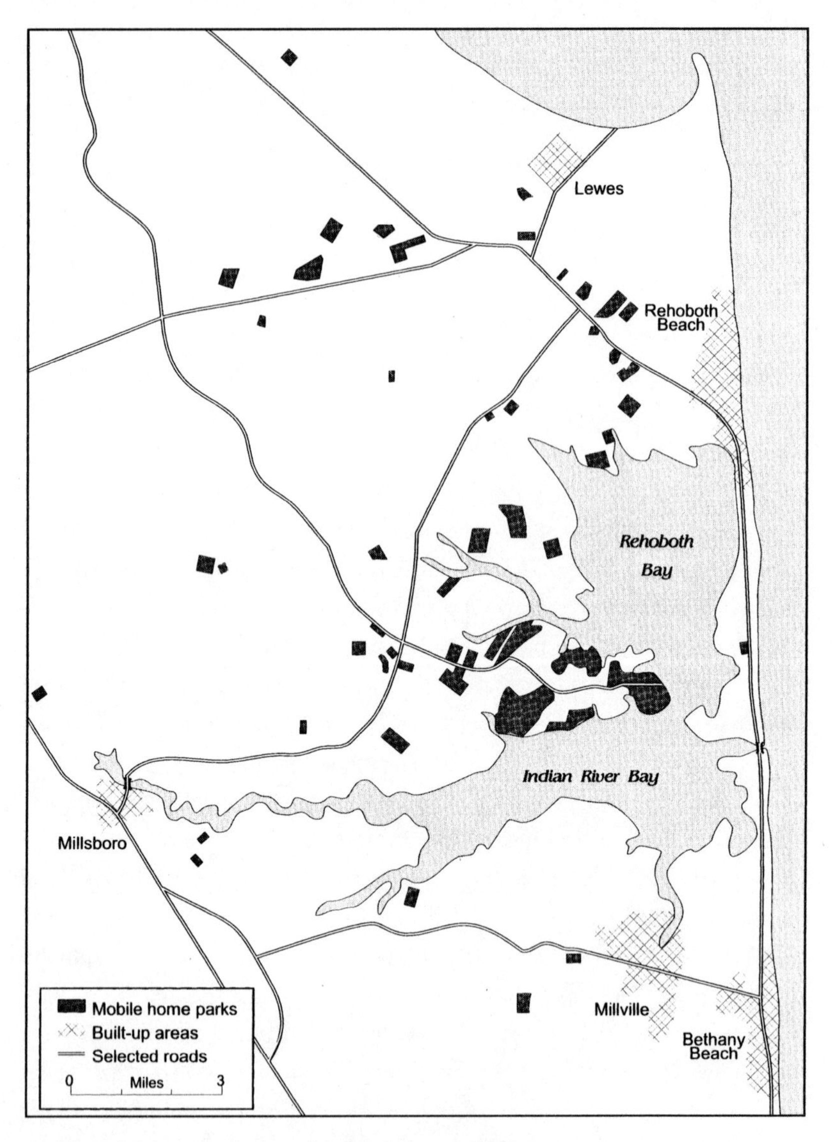

Fig. 60. Mobile home parks in southeastern Delaware, 1996

picturesque post-and-rail fences of gray weathered wood enclose each one. The elaborate gatehouses and all community buildings are painted the same gray color, which has a patina of age even when it is new. All streets are paved and signposted. The entrance signs and all street signs have the same motif of white letters on a navy blue background with blue trim, and some are adorned with the logo of a white seagull on three white pilings against a navy blue background.

Everything about the Tunnell Communities says that they have been developed by a person who loves them deeply and has lavished an enormous amount of thought and care on them. That person is Rob Tunnell, 42. Rob radiates enthusiasm about everything, and his enthusiasm is contagious. It blends wonderfully with his keen aesthetic sense.

Rob said that the Tunnell family have been in Sussex County for more than 200 years. His grandfather, James M. Tunnell, who was a lawyer in Georgetown, the county seat, served in the United States Senate from 1940 through 1946 and was a good friend of Harry Truman's. In 1932 he bought a 999-acre farm on Long Neck because he liked to raise timber, and he rented the cropland to a neighboring farmer. The farm was called Pot-Nets Farm, from crab pot and fishermen's nets, which explains the curious name of the mobile home communities. As late as 1954, when Rob was born, Long Neck was still a dead end, with only six mail stops; only six families lived on the entire peninsula, which 40 years later had more than 6,000 mobile homes.

In 1962 a friend of Rob's father asked about buying some waterfront land on which he could place a mobile home. "Dad told him that we would develop a mobile home park if he could get five other people to join him," said Rob, "and he had them all lined up in less than a week." The first advertising for Pot-Nets Mobile Home Park was a picture postcard with an oblique aerial photograph of four homes parked beside the first dredged lagoon. Rob has framed the one his mother sent to her parents in April 1963. "We now have fourteen trailers," she wrote, "and Robert is hoping to have twenty this year. . . . I wish this thing would catch on. So many people tell us it can't miss."

"Before 1981 we were unbelievably informal," Rob said. "We didn't even have a phone where people could call us. Dad was also an attorney in Georgetown, and he was constantly improving the community. For 20 years he put everything back into it, and he never took a penny out of it."

Fig. 61. A single-wide mobile home enhanced with add-ons in the Pot-Nets Mobile Home Park on Long Neck in southeastern Delaware

Fig. 62. Seems like almost everyone in Pot-Nets has a golf cart in which to putter about the community

Rob graduated from college in 1980. He had planned to become a lawyer, like his father and grandfather, "but I goofed around in college too much," he said, "and my LSAT scores were lousy. Then I got serious and took an M.A. in accountancy. I took over the parks in 1981." His uncle still owned a half-interest in the parks, and his father helped Rob to buy him out.

"Things were really spinning in the 1980s," Rob said. "In one single year we rented 160 lots, and I can remember 23 homes on the street, just sitting there waiting to be moved onto lots. In 1981 we had 1,000 lots in three parks. In 1996 we have 3,050 lots in seven parks, and we are adding more than 700 more. I employ 55 people year-round and 120 at the summer peak. We only have two and a half homes per gross acre, which gives us that nice open look. Most parks have six or more. We also have acres and acres of lagoons and nature preserves."

Half of the residents are from New Castle County, the northernmost county in Delaware; a quarter are from eastern Pennsylvania; and the rest are from Maryland and southern New Jersey. Forty percent live here year-round, and a large percentage of them are retired. The percentage who are permanent goes up a couple of points each year. Some come here and try it out for a few years to see if they like it before they move here permanently.

Visitors to the office of the Tunnell Communities, whether prospective customers or inquisitive geographers, are welcomed by a friendly receptionist who hands them a looseleaf notebook with a separate page for each home that is for sale. The standard page has two color photographs of the home and a 44-item checklist that includes external features (length, width, siding, roof, storage shed, deck, screened porch, carport, private dock), internal layout (number of bedrooms and bathrooms, living room, dining room, den, eat-in kitchen, morning room, great room, Florida room), appliances and fixtures (dishwasher, washing machine, dryer, microwave, refrigerator, heating system, air conditioning), type of lot (waterfront, waterview, lakefront, lakeview, wooded, open, interior, corner), sales price, monthly rent, and quarterly sewer charge.

In September 1996, new homes for sale in the Tunnell Communities ranged in size from 1,280 to 1,780 square feet, in price from $45,900 to $99,900, and in lot rental from $222 to $454 per month. All had two bathrooms, and most had three bedrooms. The prices of "pre-owned" homes ranged from $11,500 for a 12-by-55-foot unit with two bedrooms and one bath on a wooded lot that rented for $222 a month to $62,500 for a 28-by-64-foot unit with three bedrooms and two baths on a waterfront lot that rented for $405 a month.

The Tunnell Communities boast everything you could ask for: miles of waterfront, many lakes and ponds, private beaches and boat docks, swimming, water skiing, fishing, crabbing, nature trails, conservation areas, tennis and basketball courts, softball diamonds, playgrounds, a miniature-golf course, bike and golf cart paths, and of course that staple of all retirement areas, shuffleboard courts. Bicycles are ubiquitous, and everyone seems to have a golf cart in which to putter about.

The residents with whom we chatted were enthusiastic about the well-groomed appearance and beauty of the parks, and they told us about the dogwood blossoms in the spring and the migrating waterfowl in the fall. They liked living in a land-lease community that allows them to save money on their homes and avoid real estate taxes. They said the parks are a great place to relax, far quieter than the shore areas, and they have an excellent 24-hour security patrol system.

The residents appreciated the friendliness of all the employees of the parks, and the complete absence of high-pressure sales tactics. In fact, one of the most effective selling points of the Tunnell Communities is the enthusiasm of their residents, and Rob Tunnell has developed an Easy Money resident referral program that in 1996 awarded a resident $500 for recruiting anyone who bought a new home, $250 for recruiting anyone who bought a pre-owned home, and $150 for recruiting anyone who rented a lot.

Rob has the knack of making everyone, residents and employees alike, feel like valued members of the family. He said,

> There is a strong sense of community here. I think that mobile home parks generally have a strong sense of community, and we do everything possible to foster it. I take pride in the community. It is a first-class operation, and I see it as a partnership. I hold a public meeting for all residents once a year, and I meet with the associations six or seven times a year, whenever it is necessary.
>
> We never had paved roads, and we were lucky, because the county tore up everything when it installed a $30 million sewer system with 100 miles of pipe. Then I went to the residents and asked for a vote if they were willing to pay for hard roads, and they agreed overwhelmingly. We have 54 miles of paved roads, with 754 street signs. Those street signs alone cost me several hundred thousand dollars.
>
> At one of the meetings a resident said that we need a good golf course here. I had already been thinking about it, and it seemed like a good idea, so I am developing Baywood Greens out on State 24. It is an 18-hole championship golf course with 750 units. I want attractive units with two-car garages, and I realized that the only way I was going to get them was to have multistory units, one single-wide on top of another.
>
> I discovered that you can transport units 20 feet wide on rural roads inside the state of Delaware, so I found a local builder to produce the units for me. Each one is completely finished under cover, and then they are transported to the site. I plan to add 50 units a year at Baywood for the next 15 years, and I plan to sell them for

$50 per square foot or less. Baywood is going to cost several million dollars over 15 years, and I am going to need half of it up front.

I will also need a change in the county planning regulations. The county has a height limit of 15 feet on mobile homes. Nobody knows why, and it makes no sense. It probably goes back to the old highway height regulations. I talked to the administrator and to each of the five county commissioners, and they all agreed to raise the limit to 35 feet, although I only need 32. We still have two public hearings on the change, but I am optimistic that it will go through. All of these environmental regulations and other restrictions have increased the cost of development by more than 400 percent in the last ten years.

Other mobile home parks have followed Tunnell to Long Neck, but none are anywhere as nice, and some are so bleak, barren, and uninviting, with long tight rows of shiny metal boxes, that one scornful Long Neck resident described them as "Louisiana fishing camps." Rob said, "I think the county commissioners have decided to stick all mobile home parks on Long Neck. I guess this is where they warehouse them, but I don't think they would allow even one today if we weren't already here." All of the mobile homes on Long Neck are in land-lease parks, and only a few are single-sited.

One of the greatest assets of the Tunnell Communities is the wonderfully friendly people who work there, as we learned firsthand. While driving around, we stopped to take pictures and discovered that we had locked ourselves out of the car we had rented. We knocked on the door of the nearest home to ask for help. The owner invited us in while he called Security. We declined, because Fraser was smoking his pipe, but through the open door we saw a spacious and handsomely furnished living room.

He came back to say that Security would soon arrive, and in five minutes Mel Beecroft drove up in his squad car. He did his best to open the door, but after fruitless effort he got on his cellular phone and called a local locksmith. By this time it had started to drizzle, so he invited us to sit in the squad car with him and offered us a cup of coffee while we waited.

"I was a cop in York, Pennsylvania, for 28 years," Mel said. "I sold my house there when I retired, and I bought a mobile home in another park here. I could not afford to sell it and move when I started working here. My park is not anywhere near as nice as this one. It has fewer rules, which is both good and bad, and it has poorer maintenance. I am lucky, because I live on a good street, and I have a nice screened porch."

"I will never again live in a site-built house," he said. "You can't go by your first impressions of mobile homes. You've got to get used to them. It's just like living in a house, only a lot cheaper, and there's almost no maintenance. They do lack storage space. They have no basements, because you go down three feet here and you hit water. Everybody that can afford it puts on an addition."

Florida

The mobile home is widely perceived as a perfectly acceptable form of housing in Florida, which has more mobile homes per capita and more per square mile than any other part of the United States. In 1990 the state had 821,000 mobile homes, which comprised 13.5 percent of its total housing stock, or more than one of every eight housing units. In rural Florida more than one of every four housing units was a mobile home, and in rural areas north of Orlando the figure was closer to one of every two (Fig. 63). Mobile homes in rural areas are generally single-sited, but in built-up areas they are clustered in parks.

The mobile homes and mobile home parks of Florida are concentrated on the west coast from north of Tampa southward to Naples. The Silver Coast, as this area is called, is quite unlike the Gold Coast on the east, which is livelier and better known. The Gold Coast attracts well-to-do people from the cities of the Northeast, people who are willing to live in high-rise apartment complexes. The Silver Coast, which is lower key and lesser known, takes its name from the color of the hair of most of its residents. It attracts retired people from the Midwest, the rural Northeast, and eastern Canada, people who prefer to live in single-family homes with yards. Every sixth dwelling unit on the Silver Coast is a mobile home, and most of the mobile homes are in parks of all sizes and all price ranges, from the luxurious to the strictly utilitarian (Fig. 64).

Accurate information on the numbers and locations of mobile home parks in Florida, as elsewhere, is frustratingly hard to come by, both because tall fences or lush tropical plantings screen many from the public eye, and because individual parks vary so enormously. Some parks are as well designed, nicely landscaped, and attractive as any middle-class suburb anywhere in the United States, but a few, to be quite blunt about it, are simply crowded and unkempt slums.

Some "parks" are no more than empty lots from which the owner derives

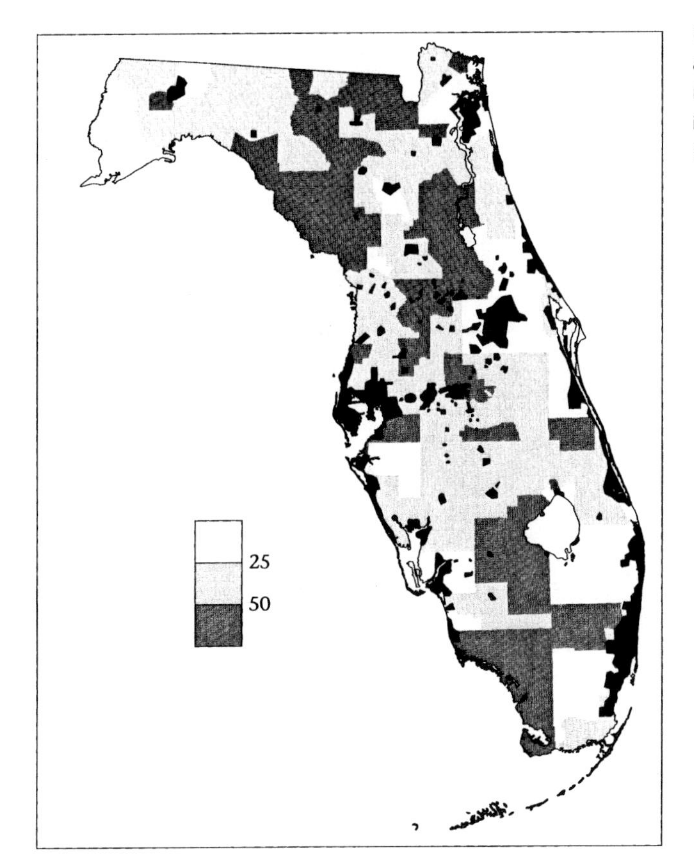

Fig. 63. Mobile homes as a percentage of all housing units outside incorporated places in Florida, 1990

a bit of income by renting parking space to the owners of three or four mobile homes. At the other extreme, in the better parks so many units have add-ons and carports that these parks are virtually indistinguishable from conventional residential developments, and it takes a really sharp eye to tell the difference.

The Zephyrhills area, 25 miles northeast of Tampa, shows why trying to obtain accurate information on the numbers and locations of mobile home parks is like trying to pin down Jell-O. Zephyrhills is a winter refuge for people who have retired on modest incomes. The center of town, where Florida 54 takes leave of U.S. 310, is dominated by a large white store with huge signs in bold red letters advertising mobile home parts and accessories such as patio covers, screen rooms, hurricane awnings, windows, doors, and aluminum siding and skirting.

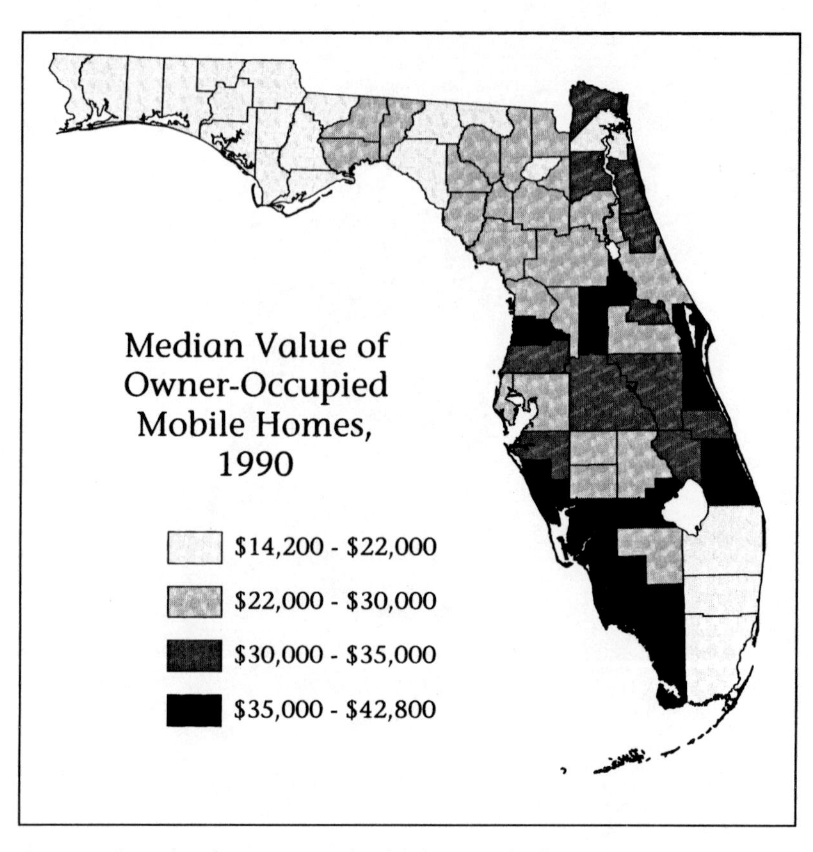

Fig. 64. Median value of owner-occupied mobile homes in Florida, 1990

On November 5, 1998, the "Welcome Back" edition of the *Zephyrhills News* published a map and directory listing 145 mobile home and RV parks but giving no indication of their size. The Zephyrhills Chamber of Commerce said that 17 parks had more than 250 spaces each, and the largest park had 994 spaces.

Some local enthusiasts boasted that the area actually had 170 parks, but the Zephyrhills (Fla.) 1:24,000 topographic quadrangle, last updated by the U.S. Geological Survey in 1987, showed only 27 "trailer parks." The topographic map did not separately identify parks that were contiguous, as some are, but the discrepancy, even given the difference in dates, suggests that many of the smaller parks had only a handful of spaces.

The 1990 Census of Housing reported that Zephyrhills had 6,000 mobile

Fig 65. Discount hardware store in Zephyrhills, Florida, that caters to mobile home residents

Fig. 66. Mobile home parks in Zephyrhills, Florida, 1999

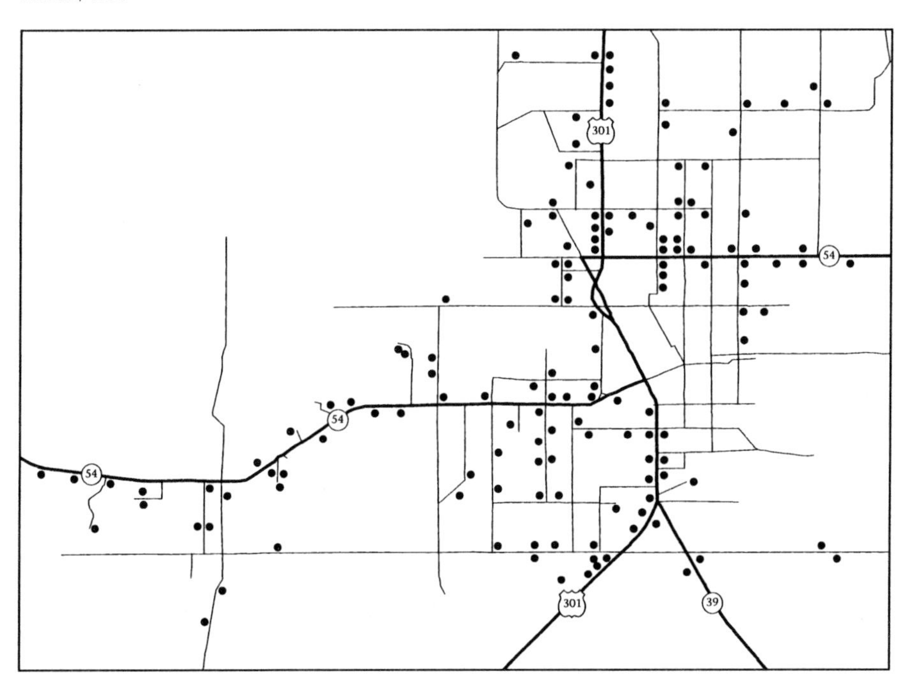

homes (50.1 percent of all housing units), and the adjacent rural area had 6,750 more (66.7 percent of all housing units). The 1990 Census of Population found that Zephyrhills had 17,300 people, and the adjacent rural area had 18,300 more. Local enthusiasts claim that the population swells to 80,000 people in winter, but 65,000 to 70,000 is probably closer to the mark, unless some housing units are awfully crowded. There can be no doubt, however, that lots of snowbirds flock to Zephyrhills in the wintertime.

The snowbirds start moving in around October 15, but the big influx is after the Thanksgiving holiday. Some drive down early and then fly home for Thanksgiving. In early November many church notice boards in Zephyrhills sport large "Welcome Back!" signs to greet the snowbirds, the local newspaper publishes its annual "Welcome Back" edition, and business owners start blowing the dust off fixtures in restaurants and other establishments that they had closed for the summer. In spring there's always a big discussion about whether to pull out before or after Easter, because people are always worried about getting caught in a late snowstorm while driving north.

One representative park in Zephyrhills had 290 lots that rented for $150 a month for the full year, and 240 lots that sold for $12,000 to $14,000 each. The park had two new units for sale at $60,000 each, but you could buy a used unit for anywhere from $3,000 to $30,000. The rental lots had septic, water, sewer, and garbage services, but lot owners had to pay for these services. Residents were allowed to have one visitor, who could stay for no more than 30 days, and they were charged $3 a night for each additional visitor.

About 60 residents lived in the park year-round, but most of the rest arrived around the first of November and headed off around the end of March. License plates were a poor indicator of their origins, because most preferred to buy cheap Florida plates, and some have elected to establish an official Florida residence because the state has no state income tax.

Two-fifths of the residents were from Michigan, one-fifth from New York, one-tenth from Canada, and the rest from all over, from states as far away as North Dakota and Maine. Michiganders were so numerous that they held an annual Michigan Picnic, to which the other residents responded by organizing an annual Everybody But Michigan picnic two weeks later.

The park's residents seemed hyperorganized, with nearly 100 officers to manage a frenetic round of social activities. In addition to the full panoply of traditional officers, it had directors for bingo, bridge, canasta, shuffleboard, arts and crafts, blood pressure checks, soup day luncheons, potluck dinners, pancake suppers, greeters, coffee pourers, doughnut servers, song leaders, ticket sales, and hall decorations.

"We all know each other" was a common theme in the residential retirement parks on the Silver Coast, which are distinctly different from rental parks. Rental parks own their units and rent them to tenants. In residential parks the residents own their units, and they rent or own the land on which their units are sited. The tenants in rental parks generally seem to have less interest in the maintenance and appearance of their units and their parks, and these parks are more likely to be junk-strewn and unsightly.

Rental parks have an adversarial atmosphere, and tenants and managers alike seem to have chips on their shoulders. The tenants suspect the manager of rent-gouging, and the manager is afraid that tenants are going to skip out without paying the rent. In one rental park we asked a woman where we could find the manager, and she snarled, "You don't want to live in this place." The managers are equally suspicious of inquisitive strangers, and we generally hit a blank wall when we tried to talk to them. Few rental parks are left, which might be just as well, because they are not friendly places, but their disappearance reduces the available stock of truly inexpensive housing.

Residential parks are quite the opposite; if anything, they are almost too friendly. Most of their residents are retired people, and they have plenty of time on their hands. Most have been coming back to the same park each winter for years, and the residents of many retirement parks are almost like extended families. The people who manage retirement parks must enjoy interacting with other people, and they are delighted to talk to inquisitive strangers, who just might be prospective residents.

Patricia Smith, 59, is the vivacious, gregarious, and tart-tongued resident manager and den mother of the K & K Mobile Home Park in Bradenton, an inexpensive park for older people who have retired on fixed incomes. The owners bought the park in 1976 as an investment property. They also own a park in Tavares and other investment property elsewhere. Pat came here in 1995, replacing a manager "who had been here since dirt was invented." The park has 175 mobiles. Fifty or so are occupied by year-round residents, and the rest are "seasonals," owned mainly by people from Michigan, New York, and Ohio.

The residents own their mobiles and rent lots for $190 to $205 a month.

The park provides water and sewer services, fire protection, and garbage collection. The average cost of a mobile is around $2,000, but in the fall of 1998 a completely furnished two-bedroom mobile with an add-on and a carport sold for $1,500, and one widow bought an old single-bedroom mobile for only $500.

"This is a great place for older folks and Canadians," Pat said. "It's an inexpensive way to live. Upkeep is easy and cheap compared to a house, and in a house they would be targets for every glib salesman that comes along." Old people want security, and the park is fairly safe. In houses they would be all alone, but in the park they look out for each other, and they get together to do jobs.

The residents are like-minded people, with a strong sense of community, and they have an active Park Tenants Association and clubhouse. Many residents are related, and in the summer up north some even get together for picnics. They have the same ailments, and they like to talk about their operations and prescriptions. "I have heard about so many knee and hip operations," said Pat, "that I think I could even perform one myself."

The park is narrow but deep. A six-foot-high woven wire fence topped with three stands of barbed wire encloses the perimeter. The entrance, at the end of the park on Fourteenth Street, is wide enough only for the white-painted cinder-block clubhouse and a small paved parking lot in front of Pat's office.

The park is a neat and tidy place. "It's their own hard-earned money," said Pat, "and they care." She has regular contests and awards prizes for the most attractive lots, and the Park Tenants Association helps her to enforce the park rules and regulations about maintenance and upkeep. Some residents are so old that they simply cannot maintain their lots, and the park has to do the work for them and charge them for it.

Pat said that she gets her own work done in the summer, when most of the residents are up north. She employs a handyman, a cleaning woman, and her son and three teenage boys part-time, but she is always happy to pull on sweats and get her own hands dirty. Many residents have excellent artisanal skills and ample time in which to exercise them. They know so many trades that there is almost always someone who can fix anything that breaks.

Screening new residents is a major responsibility and challenge, and Pat warns residents who sell their mobiles that the buyer they find might not be permitted to live in the park. At least one resident of each mobile must be 55 or over, and the park requests that all resident spouses and children be at least 45, "but there's really nothing that I can legally do about it," said Pat, "if some old geezer decides to marry an 18-year-old bimbat."

Residents may have guests, but for no more than 15 consecutive days or more than 30 days a year. Neither is considered a guest if two people are living together but not legally married, although both must be 55 or older. Some older couples elect to cohabit without benefit of clergy, because they would lose their benefits if they remarried. The residents of the park are of a generation that frowns on such behavior, but Pat said that the statutes of the state of Florida prevent her from asking to see a marriage license.

In 1999 she was a bit uneasy about the immediate future, but she thought things would improve in a few years when the members of the baby-boom generation started to retire. The average age of her current residents is 72, and "I keep a stack of sympathy cards on my desk," she said. "The kids don't want to live here when a resident dies. They clean out the place and sell it if they can, or simply turn the title over to the park. We just accept it, because it's not worth the cost of a lawsuit."

The park does not want to own any units, because they are such a headache. Pat renovates the unit and tries to sell it, but rents it if she cannot sell it. "I don't like to have to deal with smart-aleck 20-year-olds in rental units," she said. "Renters are a mess. How do you get them out? Judges don't like to evict people, and it's cheaper just to pay them to leave."

Pat is going to have to rent 12 to 14 mobiles for the next few years, but she looks forward to the time when the baby-boomers start to arrive. The older part of the park needs a lot of work, but she is starting to upgrade it. "Nobody is building mobile parks any more," she said, "only RV parks, and RVs cost too much. Some of the baby-boomers are going to need inexpensive housing, because they can't all afford luxury units."

The Six Lakes Country Club in North Fort Myers is an upscale, if not luxurious, mobile home park that laps around a private 18-hole golf course whose fairways are speckled with the six lakes that give the park its name. The paved streets that curve gently through the park are named for such illustrious golfers as Snead, Hogan, Nicklaus, and Palmer.

Johnny Johnson, a produce farmer from central Florida, developed the park in 1974. He was an avid golfer, but he knew where the money was, so he

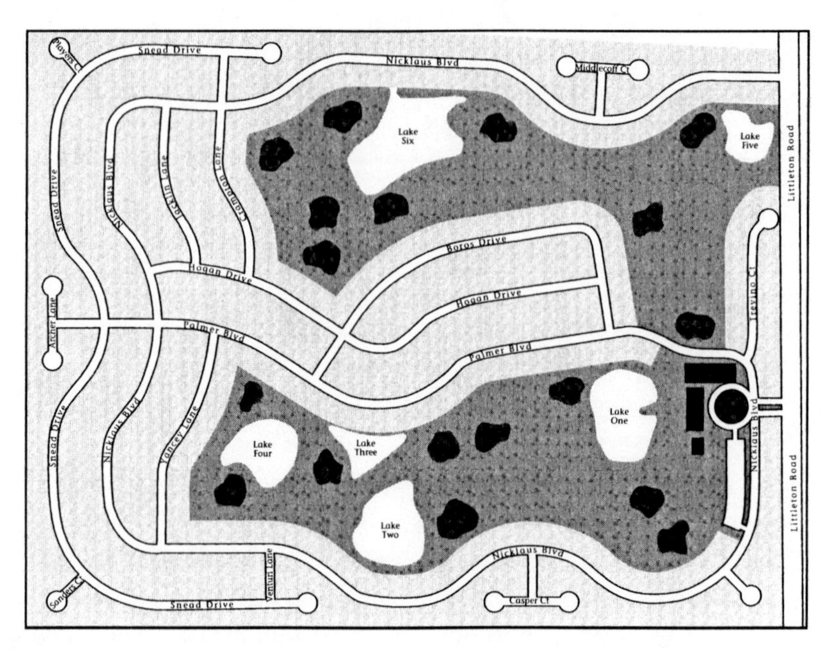

Fig. 67. Six Lakes Country Club in North Fort Myers, Florida. North is to the right.

laid out a mobile home park around his golf course. He built a handsome office and recreational hall, an attractive sandwich stand and pro shop, a swimming pool, a putting green, and several shuffleboard and tennis courts. He laid out 595 lots, mostly 60 by 80 feet, of which 157 were on the fairways. Within a few years his staff had sold all of them.

In 1984 the residents of the park formed a co-op and bought the park for $6 million, which they now say was the smartest thing they ever did. Their numbers included ample legal and professional talent to handle all details of the transaction. The co-op offered shares at $10,000 apiece to raise the money to buy the park, and expected each resident to buy one share. By 1998 the value of a share had risen to $16,180, and only eight residents did not own one. The co-op does not like to own shares, but it did hold eight in 1998. The residents own their units, and they lease their lots from the co-op. In 1998 their assessments were $2,549 for lots on the golf course and $2,223 for other lots.

A high wall encloses the park, with a gatehouse at the principal entrance and a narrow hole-in-the-wall side entrance. All gates are closed from six in the evening until six in the morning for security. The park has had three minor break-ins in the last five years, but nothing major, and it is a safe and comfortable place to live.

Six Lakes is an extraordinarily friendly place, as friendly as a Rotary Club meeting. Everyone smiles, everyone waves, everyone says hello. Everyone looks well fed and freshly scrubbed. The park manager is E. Bard Rupp, 59, genial and avuncular. He taught school in Pennsylvania for 25 years before he moved to Florida. He is professionally licensed as a manager of community associations and is essentially the city manager for a "town" of around 1,200 people. He is responsible to a board of seven directors who are elected by the residents.

The park has a separate corporation for social affairs and its own community TV channel, which it uses for announcements. It runs the Sand Wedge

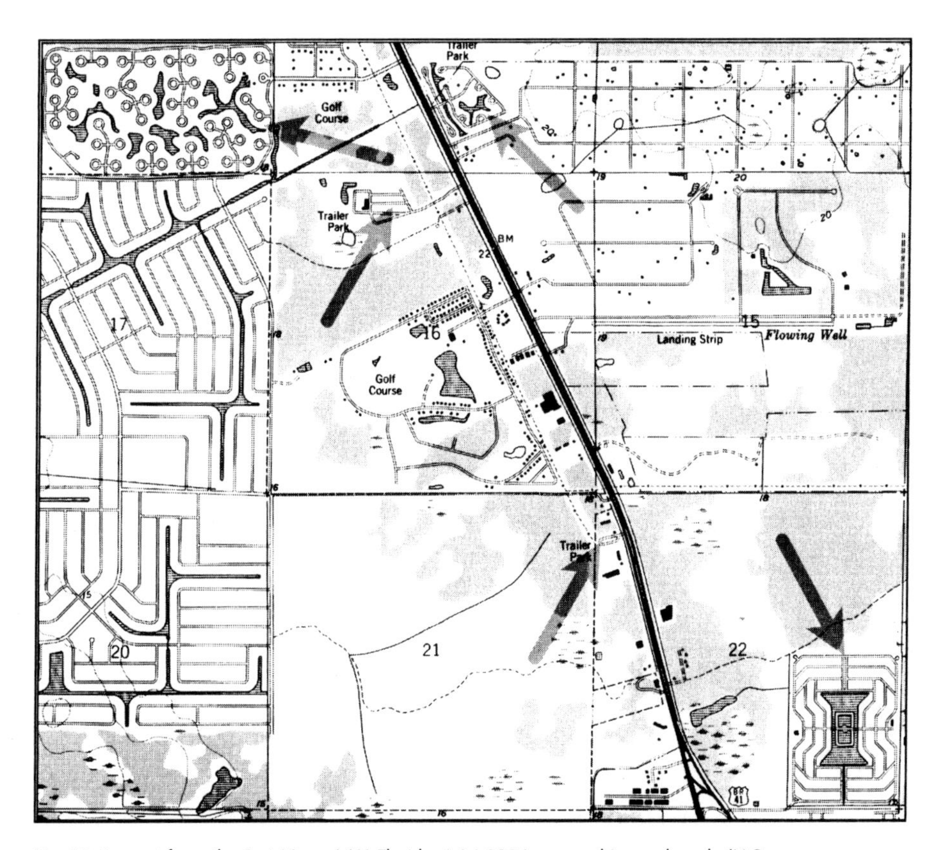

Fig. 68. Excerpt from the Fort Myers NW, Florida, 1:24,000 topographic quadrangle (U.S. Geological Survey)

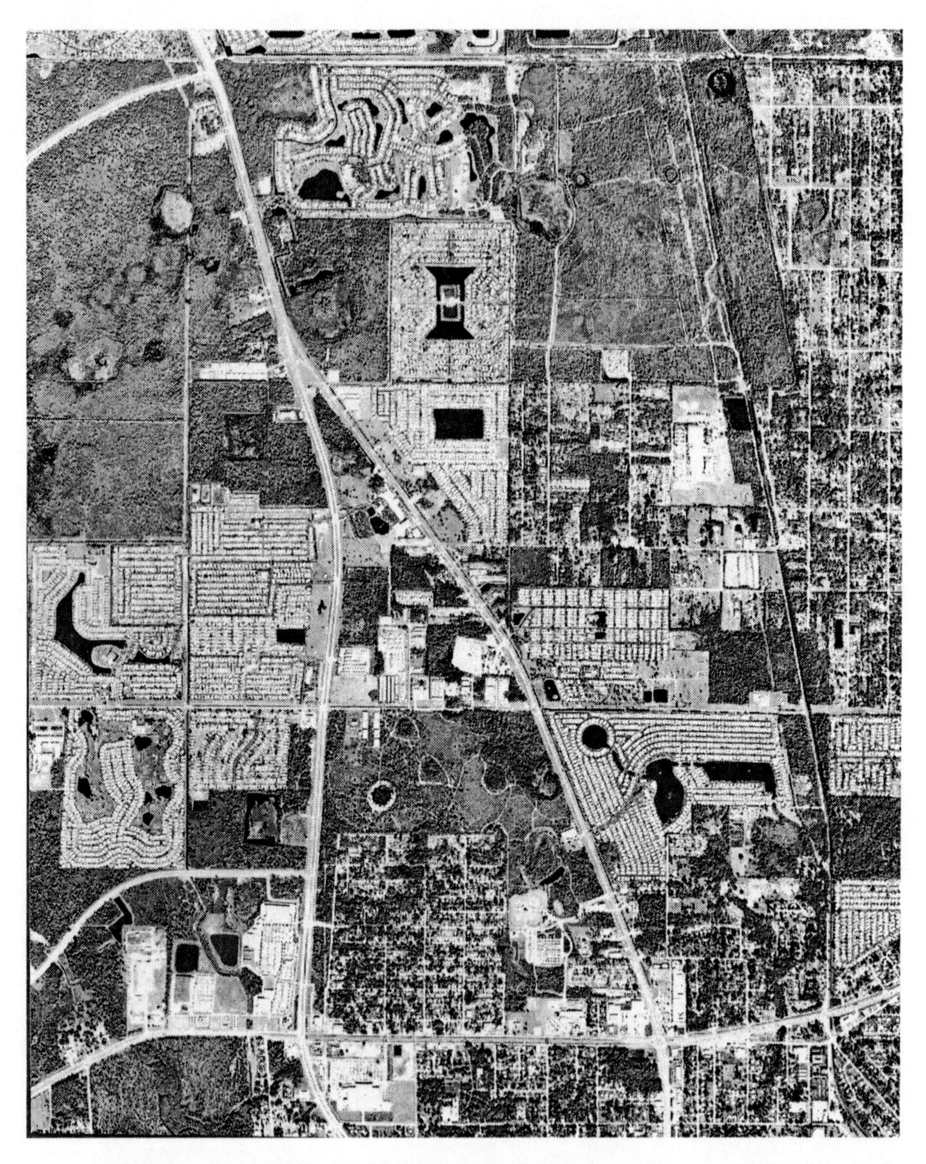

Fig. 69. Vertical aerial photograph (National Aerial Photography Program, U.S. Department of Agriculture, 11049-11, dated 01/06/99) of part of North Fort Myers, Florida. Texture differentiates mobile home parks from conventional residential areas. Six Lake Park is at the lower left, and the park in the lower right corner of Fig. 68 is in the upper center.

Shop as a nonprofit operation to keep prices low, because the residents want it that way. The park has 18 full-time and 21 part-time employees, including a golf pro and a professional greenskeeper. It provides uniforms for its employees and has a good benefits program for them.

Back in the 1970s the park was pretty far out in the country, and there were no codes or municipal regulations, but Johnson worked with the residents to keep it attractive. Residents must trim their shrubs, but the park mows all of the grass and maintains ditches and roads. It owns three water wells. Originally it had a package plant to treat sewage and used the effluent water in a computerized irrigation system on the golf course, but in 1993 it hooked on to the North Fort Myers water and sewer service system. The park provides water and utilities, for which it charges residents $30 a month.

The average resident has been in the park for 15 years or so. They come from all over, and from all income levels. Many could afford more expensive housing, but they prefer a small home that is easy to maintain. Forty percent of the residents are gone in the summer. Once they came in January, but now they come in October, and they do not leave until May.

Most of the homes in the park are 1980s-model double-wides, which probably cost an average of around $40,000. The park still has a few older single-wides, and it has already had to replace four or five, with more replacements in the offing. Bard said he got out of the real estate business a few years ago because he does not like selling, and a resident of the park, who is a licensed agent, handles all sales. Eight homes were for sale in the fall of 1998, and some of the owners were shocked by their low resale value.

Bard told me that the clubhouse building is officially certified as a hurricane shelter, and one time last year it housed 170 people. His pager went off as we were ending our conversation. The manager of the Sand Wedge Shop reported that she was out of beer, so Bard had to mount his golf cart and ride off to fetch a full keg for her.

Southern California

The transformation of mobile homes from transient travel trailers into permanent fee-simple residences has culminated in southern California. In 1980, in response to severe pressure to do something about the runaway costs of new home construction, the state legislature passed a law requiring cities

Fig. 70. Bard Rupp of Six Lakes Country Club and his trusty golf cart

Fig. 71. Unusual juxtaposition of housing types in Paso Robles, California

and counties to permit manufactured homes on any lot zoned residential if the units were placed on permanent foundations and if they had roofs and siding like the other homes in the area.

The first units to capitalize on this law were placed on individual urban infill lots in the San Francisco Bay area, but the law has also encouraged the development of large, attractively landscaped manufactured-home communities around Los Angeles. These communities are virtually indistinguishable from conventional subdivisions, except that the homes are more affordable. The homes have held their value, and the communities have served as models for the development of similar upscale communities in other parts of the country.

The new communities are designed to look and feel like well-planned conventional subdivisions. Many have only four or five homes to the acre rather than the more usual six to ten, and their lots are large enough for 16-by-80-foot single-wides or 24-by-75-foot multisectionals. Some of these communities permit only multisectionals. They have attractive new "developer series" manufactured homes with conventional siding, shingle roofs with 3/12 or 4/12 pitch, and on-site add-ons such as garages, gables, and special entryways that further enhance their appearance.

California law requires local jurisdictions to prepare housing plans to meet the full spectrum of market needs, to help young low- and moderate-income families find suitable housing and begin to build equity from owning the site as well as the home on it. Watt Industries, a traditional developer of site-built single-family subdivisions, developed Santiago Estates as a showplace manufactured-home community near Los Angeles.

The company wanted to subdivide the 168-acre site and sell individual lots and homes as fee-simple properties with conventional 30-year fixed-rate mortgages, but local officials wanted single ownership of the site, so the company developed it as a land-lease community. Buyers could finance their homes more cheaply, but the company still owns the land.

Santiago Estates was developed in three phases: 300 homes on 63 acres, then 150 homes on 30 acres, and 350 homes on 75 acres. All homes are centrally air-conditioned and have two-car garages. They range from 1,250 to 1,625 square feet and sold for $94,900 to $113,900 in 1994, when comparable site-built houses were selling for $250,000.

Some owners of older parks in southern California are upgrading them. John Curci's family, for example, has owned the Lido Mobile Home Park in Newport Beach since 1947. The park is a prize waterfront property with spectacular views, but most of the homes date from the 1950s, with flat roofs and aluminum siding, and the 30-by-35-foot lots are too small for modern units. Curci realized that he either had to close down the park or remodel it, and the only way to grow was up.

He contacted a mobile home manufacturer with the idea of building two-story mobile homes that would fit on the small lots. The manufacturer

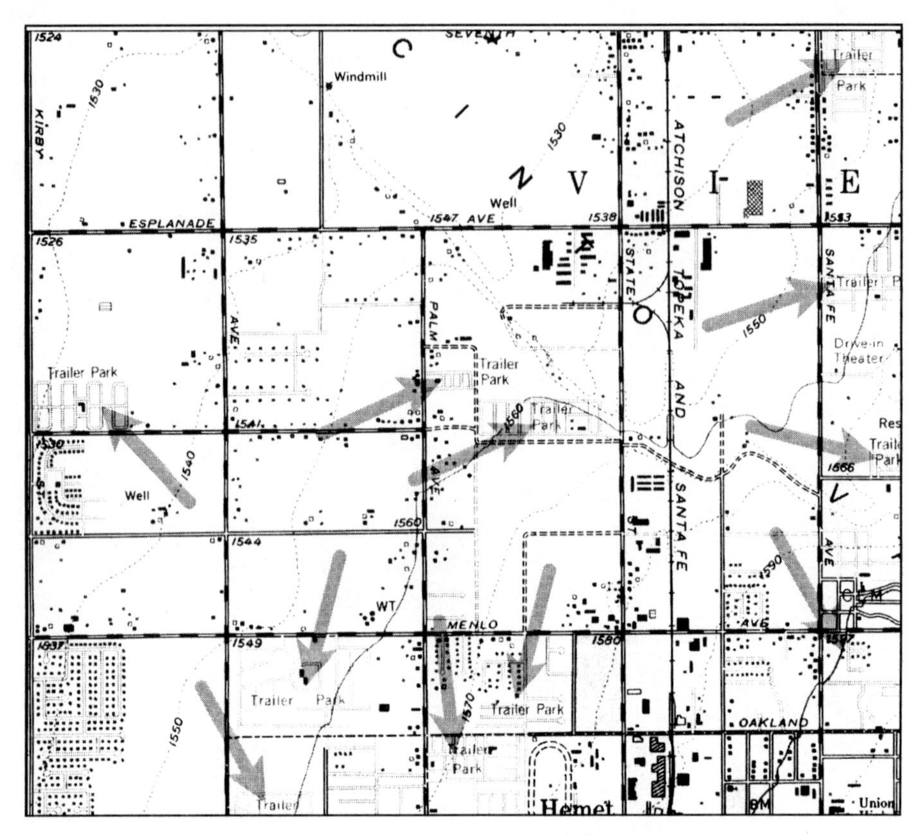

Fig. 72. Excerpt from the San Jacinto, California, 1:24,000 topographic quadrangle (U.S. Geological Survey)

developed 27-by-27-foot units with 1,000 square feet of living space, two bedrooms, and two baths. They were shipped with temporary flat roofs that were replaced by pitched roofs once they were in place. They had to be low enough to be transported under bridges and overpasses, and the manufacturer even had to lower the air pressure in the truck tires just to get them out of the factory.

Ninety of the 214 spaces in Curci's park are rented on a monthly basis, at rates of $750 to $1,850 a month, and they will be available for the new units immediately. He hopes his long-term tenants will replace their older units with the new two-story units, and he has encouraged them with favorable trade-in offers. He has already changed the name from Lido Mobile Home Park to Lido Peninsula Resort.

An $80,000 manufactured home on a lot the size of a postage stamp that rents for $20,000 a year might not seem like inexpensive housing to many Americans, but on the coast of southern California it probably qualifies as a bargain. The Pacific Coast Highway between Santa Monica and Malibu is dotted with even more expensive mobile homes that are the seaside second homes of surfers.

An increasing number of the new manufactured homes in southern California are being placed on individually owned private properties rather than in parks. Most parks are full, and few new parks are being developed, but the real reason for the shift seems to be the attractiveness of modern manufactured homes. Prospective homeowners, shocked by the cost of building a new home on site, are discovering that manufactured homes are just as nice as conventional stick-built houses, far less expensive, and virtually indistinguishable from such houses once they have been properly sited and landscaped. The new manufactured home has become an acceptable form of housing in southern California.

PART FIVE

CONCLUSION & EPILOGUE

Conclusion

Mobile home living has evolved into a common, necessary, and increasingly acceptable way of life in the six decades since the first crude camper trailer was birthed. Modern mobile homes bear scant resemblance to their camper trailer forebears, or even to their rectangular antecedents of only a few decades back, and most have been considerably modified by their owners. They are permanent additions to our national stock of affordable housing. They alleviate critical shortages in areas where housing is in short supply, serve as permanent seasonal residences in retirement areas, replace dilapidated structures, and provide permanent homes for people rural and urban throughout the nation.

The first "mobile homes" were camper trailers, often homemade, that were towed behind the family car as a better vacation home than a tent. They were unpartitioned wooden boxes without plumbing, and they had to find trailer parks with washing and toilet facilities for overnight stays. Many small towns and individual businesses near major highways developed rudimentary trailer parks to attract tourists.

During the Depression years low-income people of limited skills and limited education began to live in trailer parks permanently because they could afford no better housing. They gave trailer parks their unsavory reputation, which the media have gleefully perpetuated, as unhealthy dens of sex and violence. Many towns and cities now deem trailers and trailer parks so undesirable that they have used zoning ordinances and building codes to banish them to concealed sites on the urban fringe and beyond, where they will not offend aesthetic sensibilities.

Housing shortages near military centers during World War II forced many people to live in trailers year-round, and after the war some military, construction, mining, and other mobile families came to accept trailer living as a regular way of life. Trailers also came to be seen as inexpensive starter homes for young couples with lower levels of education, skill, and income.

After World War II, trailer manufacturers began to build wider units, which offered greater internal privacy but could not be towed safely by a family car. These larger and more livable units accelerated the transformation of the trailer from a vehicle for travel into a permanently sited residence, and the industry encouraged everyone to call them mobile homes rather than trailers. In the 1960s manufacturers started to make multisectional units, colloquially known as double-wides, two halves that could be towed separately and assembled on site.

Mobile home manufacturers have continued to make larger and better-equipped models. A modern double-wide mobile home has more floor space than the famous tract houses that were built at Levittown on Long Island in the early 1950s. Traditional single-wides are long, narrow "shoeboxes." They are short on space and storage room, and their life and contents often spill over untidily into the areas around them. Double-wides are more like conventional stick-built houses, but they are more expensive to buy and move than single-wides.

The 1974 Mobile Home Construction and Safety Standards Act directed the Department of Housing and Urban Development to create a national building code (the HUD Code) that would make mobile homes safer by reducing their vulnerability to high winds and fire. The 1980 Housing Act changed the official legal name of mobile homes to manufactured housing, but this new name is so confusing that it has not yet caught on in popular usage.

Mobile homes may be placed on lots that are owned or rented, in parks or on single sites. At the local level the distribution of single-sited mobile homes often seems random, but it can usually be explained by the convergence of three controlling factors: an owner who needs a place to put a mobile home, land that is available for the owner to buy or rent, and permissive public regulations. In urban areas single-sited mobile homes may be used to infill vacant lots. In rural areas a homeowner may let a child or relative put a mobile home in the side yard, or may sell or rent a piece of property to the owner of a mobile home.

The placement of mobile homes on single sites in rural areas is more common in the South and West than in the North. They are accepted with little apparent prejudice in remote, sparsely populated, low-income areas, where they often symbolize improved economic status. Many poor rural areas have high percentages of mobile homes in their housing stock, but a high percentage of mobile homes is not necessarily an index of poverty, because they are also a major component of the new pattern of dispersed settlement that has developed along most highways in Spersopolis, the new manufacturing belt of the Southeast.

Parks for mobile homes are ubiquitous. Virtually every city and town has utilitarian, low-income mobile-home parks at the edge of its built-up area. Older parks in what were once fringe areas have become prime sites for redevelopment as the city has expanded around them. Low-income residents have serious problems when an old park is closed and converted to more lucrative uses, because many older mobile homes cannot be moved without falling apart, and newer parks are reluctant to accept older models.

Most urban areas have chronic shortages of spaces for inexpensive starter homes in utilitarian mobile home parks. Existing parks are already filled to capacity, with long waiting lists, and few new utilitarian parks are being developed, because land on the contemporary urban fringe is too scarce and too expensive. Planners want to preserve it for open space, and they dislike designating areas for mobile home parks.

In rural areas utilitarian mobile home parks provide quick and inexpensive housing for large numbers of people near construction sites, military bases, and new industrial plants. Objections to such parks are often tinged with racism. Utilitarian parks also provide quick temporary housing for people who have been displaced by major disasters, such as floods and hurricanes. All too often these disaster parks become more permanent than was intended because their residents cannot afford to leave them.

Upscale mobile home parks have been developed in recent years because the handsome new models of mobile homes need a setting more congenial than traditional utilitarian parks provide. A few of the new upscale parks, which the developers prefer to call "resorts" or "communities," are in the

Fig. 73. Some mobile homes are truly different!

urban fringes of major cities, but most are in resort and retirement areas in Florida and California. The new upscale parks look like conventional suburban subdivisions, and retired people who are comfortably well-off use them as permanent or seasonal homes in the Sunbelt.

Living in a mobile home park, whether utilitarian or upscale, requires a special kind of gregariousness. The parks are semipublic places, because many homes are so small that residents live much of their lives outside, in each other's faces. Residents are generally neighborly folk, open and outgoing, curious about everyone else's business, and open about their own affairs. They look out for each other, sometimes perhaps to excess. A mobile home park is no place for a misanthrope or a recluse, or for those who treasure their privacy.

The majority of mobile homes sold in the United States today are multisectionals, and an increasing number are three- or even four-sectional. Some are two-story, and many are designed to have breezeways and garages attached on site. As they expand in square footage and in amenities, they increasingly appeal to middle-class homebuyers seeking economical alternatives to expensive cookie-cutter site-built houses. The new multisectionals are virtually indistinguishable from site-built ranch houses. They are finding their way onto a wider variety of sites, from traditional parks to exclusive riverfront lots and mountain hideaways.

The continuing success of mobile homes hinges in large measure not just on retaining low-income homebuyers but on attracting those of middle income. The downside is that the most affordable mobiles for low-income buyers are not new single-wides but older mobiles in rural parks, because new mobiles, at prices of $70,000 and up, are priced beyond their reach.

New or used, single or multisectional, mobile homes of all designs are one of the fastest-growing sectors of the housing market, especially in the West and South, where the majority of new homes outside major metropolitan areas are mobiles. If you are buying a new home in any rural area in the United States, the odds are that it's a mobile home. It might be in a park, or on a plot of land you own, but it's home sweet home, and it's there to stay.

Epilogue

Trailers and mobile homes, which originated as traveling vacation homes, have become affordable permanent residences for millions of Americans. That might be the end of the story, but the original travel function of trailers has now been taken over by a new breed of vehicles: recreational vehicles (RVs), camper vans, and pickup trucks with camper backs. Large RVs and camper vans may superficially resemble trailers, but they are smaller, with less than 400 square feet of space. (A standard Winnebago measures 36 by 10 feet.) Units with more than 400 square feet are classified as mobile homes, and they are not allowed in RV campgrounds, even though some mobile home parks may rent space to RVs.

RVs are self-propelled and self-contained. They are 30 to 50 feet long, 12 feet high, and 8 to 12 feet wide, and on average they cost $80,000 to $100,000. Their fuel tanks hold up to 200 gallons, but they get only 7 to 12 miles to the gallon. They have to replenish their water supply and empty their holding tanks every few days at hookups, but between hookups they can overnight at truck stops or in Wal-Mart parking lots. Often they tow small cars or boats in which their owners can make local trips when they are parked.

Camper vans and camper-back pickup trucks are primarily for camping trips, but RVs have all the comforts of home, with their own generators, central air conditioning, large-capacity water heaters, propane gas ranges and ovens, microwaves, two-door refrigerators, rear-view cameras for the driver, satellite television dishes on the roof, and large "basement" storage areas beneath the floor.

RVs spend much of their time in parks, because operating them on the highway is expensive and stressful. Parks for RVs are common throughout the Sunbelt, but the largest and best-known cluster, in the desert east of Phoenix, Arizona, has more than 40,000 spaces in more than 100 parks. Most of these parks are clean, orderly, and tranquil, with wide streets, regular street patterns, and standard lot sizes. The larger and more elaborate "resorts" are mini-cities, with sales and repair services, stores, restaurants, entertainment, and medical and other services.

The RV parks east of Phoenix are especially attractive to "snowbirds," retired people from the Midwest who flee to Phoenix to escape the northern winter. They start arriving in early November—in the desert you can tell it's autumn when the color of the license plates changes—and return to the Midwest in April to escape the scorching summer heat of the desert.

Robert C. Mings and Kevin E. McHugh have found that most snowbirds are retired white-collar middle-income couples from small towns and rural areas in the Midwest. The mild winter climate of Arizona is the initial attraction, but social ties in the RV parks quickly become important, and these people return to the same spaces in the same parks year after year. Some rent their spaces, others buy them for $12,000 to $25,000. Many could easily afford a more affluent lifestyle, but they prefer the social aspects and the camping-out feel of RV parks.

Snowbirds are gregarious folk who have large amounts of discretionary time, and the RV parks emphasize their social programs. One typical park listed no fewer than 75 activities, such as card games, dances, potluck meals, group tours, and shuffleboard tournaments. Many residents proudly display their names, home towns, states, and provinces on their RVs, which are set on "lawns" of colored gravel decorated with plastic lawn ornaments. Some parks require residents to wear large official name tags, ostensibly to allow park security to identify outsiders, but also as a get-acquainted technique.

The Phoenix area also has mobile home parks, but they are less homogeneous demographically, economically, and socially than the RV parks. They lack the recreational facilities and programs that seem essential to the RV lifestyle, and RV people consider themselves higher on the social ladder than those who live in mobile homes, perhaps because RVs cost more than most single-wides.

Most RV snowbirds "commute" between their summer homes in the

Midwest and their winter homes in the desert, but an indeterminate number have become complete nomads. They have sold their former homes, stored their belongings, bought the RV, and are free to wander. They keep in touch by e-mail and by cellular phone. In the spring they head for cool mountain campgrounds or the lush forests of the Pacific Northwest, and they return to the desert in the fall.

Early retirement and better health thanks to medical advances have also encouraged some people to take to the road as permanent nomads. Many motorcoach owners fall into this group. Motorcoaches are top-of-the-line RVs that have been luxuriously customized like tour buses for rock stars, with such touches as leather couches, jacuzzis, and fountains or electric fireplaces in their living rooms. They can cost anywhere from $250,000 to $1.5 million and up and are especially popular with retired professionals in their late 50s and early 60s.

Motorcoaches stay at plush "resorts" ("Please don't call them campgrounds!") with extra-wide streets, paved parking spaces, and full hookups for water, sewer, electricity, telephone, cable television, and anything else they need. Spaces in these resorts cost $50,000 to $150,000, and they rent for $25 to $80 a night, but people who can afford to buy motorcoaches are not about to quibble over prices where top quality is concerned. They get their mail through forwarding services and handle their finances through their brokers.

At the other extreme are the people Rene Sanchez has dubbed "geriatric gypsies," who might equally well be called the aging members of the Beat Generation. They live in anything that will move—ramshackle school buses, ancient transit buses, old delivery trucks, aging trailers and RVs—and camp free on remote areas of public land or on abandoned military bases where no one will disturb them. With friends over cards or around their campfires, they swap information about campsites good and bad, and every few weeks or so, as the spirit moves them, they drift off to another site.

BIBLIOGRAPHY

Abbott, Carl. "The Metropolitan West." In *The Twentieth Century West: Historical Interpretations,* edited by Gerald D. Nash and Richard Etulain, 71–98. Albuquerque: University of New Mexico Press, 1989.

"Annexation a Swear Word to Some." *Daily Inter Lake* (Kalispell), November 17, 1996, A1 and A3.

Bragg, Rick. "Added Game for Atlanta: Define Find a Redneck." *New York Times,* July 21, 1996, A1.

Bula, Frances. "Home Sweet (Trailer) Home." *Vancouver Sun,* November 20, 1999, A1ff.

California Department of Housing and Community Development. *Manufactured Housing for Families.* Sacramento, 1990.

Carlson, Alvar W. *The Spanish-American Homeland: Four Centuries of New Mexico's Río Arriba.* Baltimore: Johns Hopkins University Press, 1990.

Carlson, Virginia. "Local Area Profile: Elkhart-Goshen MSA, Indiana." In *Midwest Economic Report* (Federal Reserve Bank of Chicago), 10–13. 1993 third quarter.

Carroll, Jeff. "Manufactured Housing Update." *Urban Land* 56 (1997): 43–47.

Cooper, Morley. *The Trailer Book: How to Select, Equip, and Maintain a Modern Trailer Coach.* New York: Harper and Brothers, 1950.

Cowgill, Donald Olen. *Mobile Homes: A Study of Trailer Life.* Washington, D.C.: American Council on Public Affairs, 1941.

"Dream Home . . . or Nightmare." *Consumer Reports,* February 1998, 30–35.

Edgcomb, James A. "Zoning for Manufactured Housing: A Case Study in Missoula, Montana." M.Sc. thesis, University of Montana, 1988.

Evenden, L. J. "The Expansion of Domestic Space on Vancouver's North Shore." In *A Social Geography of Canada: Essays in Honour of J. Wreford Watson,* edited by G. M. Robinson, 220–44. Toronto: Dundurn Press, 1991.

Field, Thomas P. *Mobile Homes of the Kentucky and Lexington Hexagon: A Study in Areal Distribution.* Kentucky Study Series no. 5. Lexington: University of Kentucky Department of Geography, in cooperation with the Fayette County Geographical Society, 1972.

Firestone, David. "Governor's Mansion is a Triple-Wide." *New York Times,* July 12, 2000, A16.

Fitchen, Janet M. *Endangered Spaces, Enduring Places: Change, Identity, and Survival in Rural America.* Boulder, Colo.: Westview Press, 1991.

Flanagan, Barbara. "Trailer Poseurs: From Cabanas to Offices." *New York Times,* June 8, 2000, F12.

Flathead County Master Plan. Kalispell, 1987.

Flathead County Zoning Regulations. Kalispell, 1993.

Flathead River Flood Warning Map for the Flathead River Valley from Badrock Canyon to Flathead Lake. Washington, D.C.: U.S. Army Corps of Engineers, 1994.

Flathead Valley: Detail Map and Street Index. Lakeside, Mont.: dTG Maps, 1994.

Foster, Richard H., Jr. "Wartime Trailer Housing in the San Francisco Bay Area." *Geographical Review* 70, no. 3 (1980): 276–90.

French, Robert Mills, and Jeffrey K. Hadden. "An Analysis of the Distribution and Characteristics of Mobile Homes in America." *Land Economics* 41 (1965): 131–39.

Gaberlavage, George. *Issues in Manufactured Housing.* Washington, D.C.: American Association of Retired Persons, 1992.

Geisler, Charles C., and Hisayoshi Mitsuda. "Mobile-Home Growth, Regulation, and Discrimination in Upstate New York." *Rural Sociology* 52, no. 4 (1987): 532–43.

Hart, John Fraser. "Migration to the Blacktop: Population Redistribution in the South." *Landscape* 25, no. 3 (1981): 15–19.

——. "Resort Areas in Wisconsin." *Geographical Review* 74, no. 2 (1984): 192–217.

Hart, John Fraser, and John T. Morgan. "Mobile Homes." *Journal of Cultural Geography* 15, no. 2 (1995): 35–53.

——. "Spersopolis." *Southeastern Geographer* 35 (November 1995): 103–17.

Hedges, Helen E. *Key Issues in Elder Cottage Housing Opportunity (ECHO): Restrictions on Manufactured Housing.* Washington, D.C.: American Association of Retired Persons, 1991.

Hullibarger, Steve. "Manufactured Homes in Single-Family Subdivisions." *Urban Land* 55, no. 1 (1996): 42–45.

Jackson, John Brinckerhoff. "The Four Corners Country." *Landscape* 10 (1960): 20–26.

——. "The Mobile Home on the Range." In *A Sense of Time, a Sense of Place.* New Haven: Yale University Press, 1994.

Janofsky, Michael. "Texas Fugitives Fit in at Trailer Park, Neighbors Say." *New York Times,* January 24, 2001, A12.

Kaszuba, Mike. "A Suburb's Poorest Residents Move On." *Star Tribune* (Minneapolis), November 25, 1996, A1 and A8.

Ketcham, Diane. "At Pelican Lake, Birds of a Feather Roll in Style." *New York Times,* April 8, 1999, F4.

Kilborn, Peter T. "Deluxe Motor Homes Reflect Nation's Booming Economy." *New York Times,* July 4, 1999, A12.

Kunstler, James Howard. *The Geography of Nowhere.* Toronto: Simon and Schuster, 1993.

Lewis, Peirce F. "Common Houses, Cultural Spoor." *Landscape* 19 (1975): 1–22.

McHugh, Kevin E., and Robert C. Mings. "The Circle of Migration: Attachment to Place in Aging." *Annals of the Association of American Geographers* 86, no. 3 (1996): 530–50.

Marinier, Gerard. *Le caravaning: Tourisme—vacances.* Paris: Larousse, 1967.

Martin, Charles E. "The Half-House: Influences in the Creation of a New Form." In *Perspectives in Vernacular Architecture, III,* edited by Thomas Carter and Bernard L. Herman, 28–43. Columbia: University of Missouri Press, 1989.

Mings, Robert C. "Recreational Nomads in the Southwestern Sunbelt." *Journal of Cultural Geography* 4, no. 2 (1984): 86–99.

Mings, Robert C., and Kevin E. McHugh. "The RV Resort Landscape." *Journal of Cultural Geography* 10, no. 1 (1989): 35–49.

Moffat, Riley. *Population History of Western U.S. Cities and Towns, 1850–1990.* Lanham, Md.: Scarecrow Press, 1996.

Montana Department of Commerce. *1993 State of Montana Comprehensive Housing Affordability Strategy (CHAS): Final Report to HUD.* Helena, 1993.

Murphy, Kim. "In Sammyville, Folks Stick to Their Guns." *Star Tribune* (Minneapolis), December 17, 1999, A33.

Nieves, Evelyn. "Hidden Away, Trailer Park Watches as Silicon Valley Drives By." *New York Times,* November 8, 2000, A16.

Nugent, Walter. "The People of the West since 1890." In *The Twentieth-Century West: Historical Interpretations,* edited by Gerald D. Nash and Richard Etulain, 35–70. Albuquerque: University of New Mexico Press, 1989.

Nutt-Powell, Thomas E. *Manufactured Homes: Making Sense of a Housing Opportunity.* Boston: Auburn House, 1982.

Personick, Martin E., and Judy R. Daley. "Profiles in Safety and Health: Work Hazards of Mobile Homes." *Monthly Labor Review* 112, no. 7 (1989): 15–20.

Quick Facts: The Latest Trends and Information on the Manufactured Housing Industry. Arlington, Va.: Manufactured Housing Institute, 1995–96.

Rhodes, Michelle Jo Ann. *The Manufactured Home: House Form and Region.* M.A. thesis, Department of Geography, Simon Fraser University, 1998.

Sanchez, Rene. "When Retirement Is a Road Trip." *Washington Post,* February 10, 1999, A10 and A16.

Sedan, Paul S. *The Factory-Crafted House: New Visions of Affordable Housing Design.* Old Saybrook, Conn.: Globe Pequot Press, 1992.

Sheldon, Johnathan, and Andrea Simpson. *Manufactured Housing Park Tenants:*

Shifting the Balance of Power, a Model State Statute. Washington, D.C.: American Association of Retired Persons, 1991.

Speelman, Jo Ann. "City Annexations Often Controversial." *Daily Inter Lake* (Kalispell) November 20, 1996, A5.

Stull, Donald L., Janet E. Benson, Michael J. Broadway, Arthur L. Campa, Ken C. Erickson, and Mark A. Grey. *Changing Relations: Newcomers and Established Residents in Garden City, Kansas.* Report no. 172. Lawrence: University of Kansas Institute for Public Policy and Business Research, 1990.

Tauxe, Caroline S. *Farms, Mines, and Main Street: Uneven Development in a Dakota County.* Philadelphia: Temple University Press, 1993.

"Trailer Test." *Time,* November 23, 1936, 67–68.

"200,000 Trailers." *Fortune* 15, no. 3 (1937), 105ff.

United States Bureau of the Census. *1990 Census of Population and Housing, Summary Population and Housing Characteristics.* CPH-1 series, state tables 7 and 8, various dates.

United States Department of Housing and Urban Development. *Manufactured Housing: A HUD User Resource Guide.* Washington, D.C., 1993.

Wallis, Allan D. "Assimilation and Accommodation of a Housing Innovation: A Case Study Approach of the House Trailer." In *The Meaning and Use of Housing,* edited by Ernesto G. Arias, 425–41. Avebury, Mass.: Aldershot, 1993.

——. *Wheel Estate: The Rise and Decline of Mobile Homes.* New York: Oxford University Press, 1991.

Warner, Kate. *Social and Economic Impacts of Mobile Home Parks.* Ann Arbor: University of Michigan Architecture and Planning Research Laboratory, 1986.

Wheeler, James O., Gale Callaghan, and Gordon Brewer. "Locational Factors in the Growth of Mobile Home Manufacturing in the Southeastern United States." *Southeastern Geographer* 13, no. 2 (1973): 92–104.

Wolinsky, Cary. "America's Largest Parking Lot 85346." *National Geographic* 199, no. 1 (2001): 124–27.

INDEX

About the Authors

John Fraser Hart was born in 1924 in Staunton, Virginia, and was raised in Virginia and Atlanta. He received an M.A. and a Ph.D. in geography at Northwestern University. His academic honors include an American Geographical Society Medal in 2001, the first Lifetime Achievement Award from the Southeastern Division of the Association of American Geographers in 1987, a John Simon Guggenheim Memorial Foundation Fellowship in 1982–83, the presidency of the Association of American Geographers in 1979–80, the editorship of the *Annals of the Association of American Geographers* from 1970 to 1975, and an Award for the Teaching of Geography, College Level, from the National Council for Geographic Education in 1971. He is the author and editor of ten books, most recently *The Rural Landscape* from Johns Hopkins (1998). For his 1991 book *The Land that Feeds Us* he received the 1992 Geographic Society of Chicago Publication Award as well as the 1992 Association of American Geographers' John Brinkerhoff Jackson Prize. Currently, he is a professor of geography at the University of Minnesota.

Native North Carolinian *John T. Morgan* grew up working on his family's tobacco farm. After serving as an army Spanish linguist he obtained a Ph.D. in geography from the University of Tennessee, specializing in rural geography and folk architecture. Since 1986, Morgan has been on the faculty of Emory & Henry College, where he serves as chair of both the social sciences division and the department of geography. His scholarly endeavors have focused primarily on the material culture of the Upland South. He is the author of *The Log House in East Tennessee* (1990).

Michelle Rhodes, a native of northern Illinois, is currently completing a Ph.D. in geography at Simon Fraser University. The focus of her dissertation is the Montana-Alberta-Saskatchewan borderlands. Her interest in manufactured housing stems from research conducted in Flathead Valley, Montana, for her masters in geography at SFU.

Related Books in the Series

Across This Land: A Regional Geography of the United States and Canada
John C. Hudson

Along the Ohio
Andrew Borowiec

America's Original GI Town: Park Forest, Illinois
Gregory C. Randall

Belonging to the West
Eric L. Paddock

The Cotton Plantation South since the Civil War
Charles S. Aiken

Delta Sugar: Louisiana's Vanishing Plantation Landscape
John B. Rehder

Entrepreneurial Vernacular: Developers' Subdivisions in the 1920s
Carolyn S. Loeb

Everyday Architecture of the Mid-Atlantic: Looking at Buildings and Landscapes
Gabrielle M. Lanier and Bernard L. Herman

Historic American Towns along the Atlantic Coast
Warren Boeschenstein

John Nolen and Mariemont: Building a New Town in Ohio
Millard F. Rogers, Jr.

Let the Cowboy Ride: Cattle Ranching in the American West
Paul F. Starrs

Measure of Emptiness: Grain Elevators in the American Landscape
Frank Gohlke, with a concluding essay by John C. Hudson

The New American Village
Bob Thall

The New England Village
Joseph S. Wood

The Pennsylvania Barn: Its Origin, Evolution, and Distribution in North America
Robert F. Ensminger

Silent Screens: The Decline and Transformation of the American Movie Theater
Michael Putnam, with an introductory essay by Robert Sklar